Fog-Enabled Intelligent IoT Systems

Yang Yang • Xiliang Luo • Xiaoli Chu
Ming-Tuo Zhou

Fog-Enabled Intelligent IoT Systems

Springer

Yang Yang
ShanghaiTech University
Shanghai Institute of Fog Computing
Technology (SHIFT)
School of Information Science and
Technology
Shanghai, China

Xiliang Luo
ShanghaiTech University
Shanghai Institute of Fog Computing
Technology (SHIFT)
School of Information Science and
Technology
Shanghai, China

Xiaoli Chu
Department of Electronic & Electrical
Engineering
University of Sheffield
Sheffield, UK

Ming-Tuo Zhou
Chinese Academy of Sciences
Shanghai Institute of Microsystem and
Information Technology
Shanghai, China

ISBN 978-3-030-23187-3 ISBN 978-3-030-23185-9 (eBook)
https://doi.org/10.1007/978-3-030-23185-9

This Springer imprint is published by the registered company Springer Nature Switzerland AG.
The registered company address is: Gewerbestrasse 11, 6330 Cham, Switzerland

To our families—Yang Yang, Xiliang Luo,
Xiaoli Chu, Ming-Tuo Zhou

Preface

This book is focused on the potentials, opportunities, and challenges of fog computing in developing and deploying intelligent IoT systems and services. Specifically, Chap. 1 reviews key IoT technologies and several applications, which include not only simple data sensing, collection, and representation but also complex information extraction and behavior analysis. As 5G mobile networks are beginning to be commercially deployed worldwide, intelligent IoT applications and services are getting more and more important and popular in different business sectors and industrial domains, thanks to more communication bandwidth, better data quality, faster data rate, denser network connectivity, lower transmission latency, and higher system reliability.

Chapter 2 introduces the architecture and key enabling technologies of fog computing, as well as its latest development in standardization bodies and industrial consortium. As the bridge connecting the cloud and things, fog computing plays a crucial role in identifying, integrating, managing, and utilizing multitier computing, communication, and storage resources in different IoT systems. Together with feasible AI algorithms, fog nodes can combine various local/regional micro-services and orchestrate more intelligent applications and services with different user preferences and stringent performance requirements. For example, autonomous driving and intelligent manufacturing require high security in data transmission and storage, very low latency in data processing and decision-making, and super-high reliability in network connectivity and service provisioning. Furthermore, the challenges of developing more sophisticated services across multiple domains are discussed.

Chapter 3 proposes an analytical framework for general Multi-Task Multi-Helper (MTMH) fog networks and application scenarios, such as multi-robot systems, wireless communication networks, intelligent transportation systems, and smart home. Specifically, a comprehensive system model consisting of network architecture, wireless channels, communication and computing models, and task types is developed for a MTMH fog network. Based on different game theories, the fundamental problems of computation offloading are formulated and analyzed for non-splittable and splittable tasks, respectively. Accordingly, two efficient algorithms are designed and fully evaluated under different performance metrics.

Chapters 4–7 choose different intelligent IoT applications as real examples to demonstrate the architecture, functions, technologies, and advantages of our corresponding fog-enabled solutions. In particular, Chap. 4 introduces fog-enabled solutions for robot SLAM, multi-robot smart factory, and multi-robot fleet formation applications, which require large local computing power for timely constructing the map of a working environment, calculating multiple robots' exact positions, and tracking their movement postures and orientations. Through a high-speed wireless network, massive data and images collected by on-board and local sensors are transmitted from the robots and intelligent infrastructure to nearby fog nodes, where intelligent data processing algorithms are responsible for analyzing valuable information and deriving the results in real time.

Chapter 5 introduces fog-enabled self-optimization techniques for wireless communication networks, which provide end users a series of new functions and capabilities such as distributed local caching, collaborative radio signal processing, scalable service architecture, computation offloading, on-demand mobility management, and cooperative resource allocation at base stations and within the network.

Chapter 6 introduces fog-enabled intelligent transportation systems, which support autonomous driving, cooperative driving, and shared vehicles for improving user experience, traffic efficiency, and road safety. Our solution utilizes distributed computing, storage, and communication resources in vehicle nodes, roadside stations, and advanced wireless networks to realize the potential benefits of fog computing.

Chapter 7 introduces fog-enabled smart home and user behavior recognition applications, which enhance everyone's home environment, personalized services, quality of life, convenience, and safety, especially for the aging and disabled population who need a comprehensive set of interactive technologies, models, and algorithms for better assisted living experiences. Higher recognition accuracy can be achieved when millimeter wave communication devices, multitier fog nodes, and advanced algorithms are widely deployed in our buildings and neighborhood.

Chapter 8 concludes this book and suggests several key research topics for further investigations.

July 2019

Yang Yang
Xiliang Luo
Xiaoli Chu
Ming-Tuo Zhou

Acknowledgements

We would like to thank all the people who have made contributions to this book. In particular, we want to acknowledge the enormous helps from Dr. Kunlun Wang, Dr. Yang Li, Mr. Zening Liu, Mr. Penghao Cai, and Ms. Youyu Tan at School of Information Science and Technology, ShanghaiTech University, China; Dr. Wuxiong Zhang, Mr. Weidong Fang, Mr. Guofeng Shen, Dr. Yang Liu, and Dr. Shuang Zhao at Shanghai Institute of Microsystem and Information Technology, Chinese Academy of Sciences; Dr. Haijun Zhang at the University of Science and Technology Beijing, China; Dr. Jianbo Du at Xi'an University of Posts and Telecommunications, China; Mr. Jiaqi Wang at the University of Sheffield, UK; and Dr. Jiong Jin at School of Software and Electrical Engineering, Swinburne University of Technology, Australia.

Furthermore, we would like to express our gratitude to our colleagues for their kind support and encouragement.

Contents

1 IoT Technologies and Applications 1
1.1 Introduction 1
1.2 Standards and Technologies 4
1.2.1 Radio Frequency Identification 5
1.2.2 Near Field Communication 9
1.2.3 ZigBee 11
1.2.4 LoRa 13
1.2.5 Sigfox 16
1.2.6 NB-IoT 18
1.2.7 Web of Things 21
1.3 Typical Applications 23
1.3.1 Environmental Monitoring 23
1.3.2 Infrastructure Health Monitoring 25
1.3.3 Public Safety Surveillance 26
1.3.4 Smart City 28
1.3.5 Smart Manufacturing 29
1.3.6 Intelligent Transportation System 31
1.4 New Challenges 32
1.5 Conclusion 35
References 36

2 Fog Computing Architecture and Technologies 39
2.1 Introduction 39
2.2 Fog Computing Architecture 42
2.2.1 Reference Architecture 42
2.2.2 Multi-Tier Computing Network 45
2.3 Fog Computing Technologies 48
2.3.1 Networking Technologies 48
2.3.2 Computing Technologies 49
2.3.3 Storage Technologies 49
2.3.4 Virtualization Technologies 50
2.3.5 Artificial Intelligence 51

2.4 Applications and Challenges ... 52
2.4.1 Collaborative Robot System ... 52
2.4.2 Wireless Communication Network ... 54
2.4.3 Intelligent Transportation System ... 56
2.4.4 Smart Home ... 58
2.5 Conclusion ... 59
References ... 60

3 Analytical Framework for Multi-Task Multi-Helper Fog Networks ... 61
3.1 Introduction ... 61
3.2 System Architecture and Analytical Models ... 65
3.2.1 Multi-Task Multi-Helper Fog Networks ... 65
3.2.2 Non-Splittable Task and Paired Offloading ... 68
3.2.3 Splittable Task and Parallel Offloading ... 69
3.3 Theoretical Preliminaries ... 70
3.3.1 Potential Game ... 70
3.3.2 Generalized Nash Equilibrium Problem ... 73
3.3.3 Matching Theory ... 73
3.4 Game Formulation and Algorithm Design ... 75
3.4.1 Paired Offloading of Multiple Tasks ... 75
3.4.2 Parallel Offloading of Splittable Tasks ... 79
3.5 Performance Evaluation ... 87
3.5.1 Price of Anarchy ... 87
3.5.2 System Average Delay ... 89
3.5.3 Number of Beneficial TNs ... 93
3.5.4 Convergence ... 95
3.6 Conclusion ... 96
References ... 96

4 Fog-Enabled Multi-Robot System ... 99
4.1 Introduction ... 99
4.2 Simultaneous Location and Mapping ... 100
4.2.1 System Architecture ... 100
4.2.2 Technical Challenges ... 101
4.2.3 Fog-Enabled Solution ... 102
4.3 Multi-Robot Smart Factory ... 109
4.3.1 System Architecture ... 110
4.3.2 Warehouse Management ... 113
4.3.3 Emergency Management ... 119
4.4 Multi-Robot Fleet Formation ... 125
4.4.1 System Architecture ... 125
4.4.2 Fog-Enabled Solution ... 127
4.5 Conclusion ... 130
References ... 130

5 Fog-Enabled Wireless Communication Networks 133
5.1 Introduction 135
5.1.1 Centralized SON 135
5.1.2 Distributed SON 136
5.1.3 Hybrid SON 137
5.2 Self-Optimization of Mobility Management 138
5.2.1 Mobility Load Balancing 138
5.2.2 Mobility Management and Procedures 139
5.2.3 Mobility Robustness and Handover Optimization 142
5.3 Self-Coordination of Inter-Cell Interference 143
5.3.1 Interference-Aware Radio Resource Allocation 144
5.4 Self-Optimization of Coverage and Capacity 146
5.4.1 Deep-Learning Enabled Coverage and Capacity Optimization 147
5.5 Self-Optimization of Network Resources 148
5.5.1 Network Edge Caching 149
5.5.2 Computation Offloading 150
5.5.3 Joint Optimization of Computation Offloading and Resource Allocation 151
5.6 Conclusion 158
References 159

6 Fog-Enabled Intelligent Transportation System 163
6.1 Introduction 163
6.2 Intelligent Transportation System 165
6.2.1 Architecture and Key Components 165
6.2.2 Vehicle Categories and Requirements 167
6.3 Current Solutions and Technical Challenges 168
6.3.1 On-Vehicle Computing 168
6.3.2 Communication Network 170
6.4 Fog-Enabled Solution 172
6.4.1 Architecture and Key Technologies 173
6.4.2 Virtualized and Intensive Vehicle Stations 177
6.4.3 Distributed Resources in Communication Networks 178
6.4.4 Use Cases and Application Scenarios 180
6.5 Conclusion 183
References 183

7 Fog-Enabled Smart Home and User Behavior Recognition 185
7.1 Introduction 185
7.2 User Behavior Recognition for Smart Home 187
7.3 Passive Approach 189
7.3.1 Signal Detection 189
7.3.2 Wi-Fi Based Solution for Behavior Recognition 192
7.3.3 Massive MIMO Based Behavior Detection and Ranging 194

7.4 Active Approach 198
7.4.1 Millimeter Wave Radar Based Behavior Detection 198
7.4.2 Acoustics Based Localization and Motion Tracking 204
7.5 Conclusion 208
References 209

8 Conclusions 211
References 212

Index 213

Acronyms

3G	Third generation
3GPP	Third-Generation Partnership Project
4G	Fourth generation
5G	Fifth generation
AD	Autonomous driving
ADAS	Automatic driver assistant services
ADD-PS	Angle-delay-Doppler power spectrum
AGV	Automated guided vehicle
AI	Artificial intelligence
AIDC	Automated identification and data capture
AoA	Angle of arrival
AoD	Angle of departure
AP	Access point
API	Application programming interface
AR	Augmented reality
AWGN	Additive white Gaussian noise
BBU	Baseband unit
BS	Base station
CAN	Controller area network
CCo	Coverage and capacity optimization
CEAL	Cognition, efficiency, agility, and latency
CEN	European Committee for Standardization
CENELEC	European Committee for Electrotechnical Standardization
CFR	Channel frequency response
CIoT	Cellular internet of things
CIR	Channel impulse response
CNN	Convolutional neural network
CoMP	Coordinated multiple points
COP	Common operating picture
CPU	Central processing unit
CRAN	Cloud radio access network

CSI	Channel state information
CSMA	Carrier sense multiple access
CSS	Chirp spread spectrum
D2D	Device-to-device
DAG	Directed acyclic graph
DSRC	Dedicated short-range communication
DTW	Dynamic time warping
DWT	Discrete wavelet transform
E2E	End-to-end
ECU	Electronic control units
eMTC	Enhanced machine-type communication
eNB	eNodeB
EPC	Electronic product code
EPS	Evolved packet system
ETSI	European Telecommunication Standards Institute
FCC	Federal communications commission
FDMA	Frequency division multiple access
FFT	Fast Fourier transform
FIR	Finite impulse response
FN	Fog nodes
FOTA	Firmware-over-the-air
FPGA	Field programmable gate arrays
GNEP	Generalized Nash equilibrium problem
GPP	General purpose processor
GPRS	General packet radio service
GPS	Global positioning system
GPU	Graphics processing unit
HetNets	Heterogeneous networks
HII	High interference indicator
HMM	Hidden Markov model
ICIC	Inter-cell interference coordination
ICT	Information and communication technology
IEC	International electro-technical commission
IEEE	Institute of Electrical and Electronics Engineers
IMU	Inertial measurement unit
IoT	Internet of Things
ISG	Industry specification group
ISM	Industrial, scientific, and medical
ISO	International standard organization
ITS	Intelligent transport system
ITU	International Telecommunication Union
KNN	K-nearest neighbor
KPI	Key performance indicator
LIN	Local interconnect network
LoRA	Long range radio

LOS	Light-of-sight
LPWAN	Low-power wide area network
LSTM	Long short-term memory
LTE	Long-term evolution
LTN	Low throughput network
M2M	Machine to machine/man
MAC	Medium access control
MEC	Mobile/multi-access edge computing
MIMO	Multiple-input multiple-output
MLB	Mobility load balancing
MOPSO	Multiobjective particle swarm optimization
MOST	Media oriented systems transport
NB-IoT	Narrowband IoT
NE	Nash equilibrium
NFC	Near field communication
NFV	Network function virtualization
NMS	Network management system
NOMA	Non-orthogonal multiple access
NR	New radio
NSGA	Non-dominated sorting genetic algorithm
OFDM	Orthogonal frequency-division multiplexing
OI	Overload indicator
OMC	Operation and maintenance center
PAES	Pareto archived evolution strategy
PCA	Principal component analysis
PGW	Packet data network
PoA	Price of anarchy
QoE	Quality of experience
QoS	Quality of service
R2V	RSU-to-vehicle
RAID	Redundant array of inexpensive disks
RAN	Radio access network
RAS	Reliability, availability, and serviceability
RAT	Radio access technology
RF	Radio frequency
RFI	Radio frequency interface
RFID	Radio frequency identification
RNN	Recurrent neural network
RNTP	Narrowband transmit power
ROS	Robot operation system
RRH	Remote radio head
RSS	Received signal strength
RSU	Roadside unit
SCEF	Service capability exposure function
SDN	Software-defined networks

SDN-eNB	Software-defined eNodeB
SDN-EPC	software-defined EPC
SDWN	Software-defined wireless networking
SGW	Serving gateway
SINR	Signal to interference plus noise ratio
SLA	Service-level agreement
SLAM	Simultaneous localization and mapping
SON	Self-organizing networks
STFT	Short-time Fourier transform
SVM	Support vector machine
TDMA	Time division multiple access
TSN	Time sensitive networking
URLLC	Ultra-reliable and low latency communications
V2I	Vehicle-to-infrastructure
V2V	Vehicle-to-vehicle
VANET	Vehicular ad hoc network
VM	Virtual machines
VNF	Virtual network function
vSLAM	Visual SLAM
W3C	World wide web consortium
WAN	Wide area network
WLAN	Wireless local area network
WWW	World wide web
WWAN	Wireless wide area network
WoT	Web of things
WPAN	Wireless personal area networks
WSN	Wireless sensor network

Chapter 1
IoT Technologies and Applications

1.1 Introduction

As a new dimension to the world of Information and Communication Technologies (ICTs), the concept of Internet of Things (IoT) aims at providing "connectivity for anything," "by embedding short-range mobile transceivers into a wide array of additional gadgets and everyday items, enabling new forms of communication between people and things, and between things themselves," according to the seventh International Telecommunication Union (ITU) Internet Reports 2005 [1]. In recent years, different IoT technologies and standards have been actively developed for different industrial sectors and application scenarios, such as smart city, intelligent transportation system, safety and security, intelligent agriculture, environmental monitoring, smart factory, intelligent manufacturing, smart home, and healthcare. Gradually, specific IoT-centered business models and value chains have become matured, consolidated, and popular, these IoT applications are effectively accelerating the digitalization and informationization progresses of various traditional industries [2]. As a result, they have generated tremendous economic and social benefits to our society, as well as huge impacts on everyone's daily life.

Sensors, machines, and user devices are the "things," which are usually equipped with limited energy resources and simple sensing, computing, communication, motion, and control capabilities. By using a large number of sensors in the field, a typical IoT system can effectively expand its service coverage and capability in sensing, collecting, and processing different types of raw data obtained from the physical world. Most redundant data with low values will be aggregated or filtered for saving scarce communication resources, i.e., bandwidth, storage, and energy. Selected data with potential values, e.g., characteristics of unexpected events, will be transmitted from different sites through multi-hop communication networks to a centralized computing facility, such as data center, for further in-depth investigation.

Y. Yang et al., *Fog-Enabled Intelligent IoT Systems*,
https://doi.org/10.1007/978-3-030-23185-9_1

New information will be extracted, or new events will be discovered, from this more comprehensive analysis of massive data from a much larger area across multiple sites and locations.

In early days, IoT systems are usually developed according to rigid rules or requirements. The main purpose is to improve the perception of the physical world, as well as the efficiency of data collection and analysis, in different IoT applications such as environmental monitoring and emergency management. As a well-known application-driven IoT architecture, the ISO/IEC 30141-IoT Reference Architecture is often adopted in system designs and service deployments [2]. Firstly, data acquisition involves all kinds of sensors, such as RFID, MEMS, bar code, and video camera. However, due to dynamic application scenarios and environments, the key function and challenge for IoT systems is high-quality data collection (transmission) through wireless ad hoc networks. Many wireless access and networking technologies have been developed for ensuring timely and reliable connectivity and transmission for data collection at low cost and low energy consumption [3–5]. In addition to the existing standards for mobile communications, the Internet, and broadcasting networks, a series of wireless communication technologies have been developed for supporting short-distance data transmissions in various IoT application scenarios, e.g., RFID, Wi-Fi, NFC, ZigBee, LoRa, and Sigfox [6–11]. Finally, by collaboratively analyzing data from multiple sensors in different areas, a more comprehensive perception of the actual environment and a timely understanding of the exact situation will be achieved, thus enabling better decision making and performance optimization for particular industrial operations.

Nowadays, a series of advanced technologies on smart sensing, pervasive computing, wireless communication, pattern recognition, and behavior analysis have been widely applied and integrated for supporting more and more sophisticated IoT applications, such as intelligent transportation system and intelligent manufacturing. Such complex applications can greatly improve system automation and efficiency in massive data analysis and task execution. To achieve this goal, the key function and challenge for IoT systems is accurate information extraction, which heavily depends on domain-specific knowledge, valuable experience, and technical know-hows contributed by real experts and field workers. In order to make IoT systems more accessible and deployable, the fourth-generation (4G) and fifth-generation (5G) mobile communication standards have specified several important IoT application scenarios, i.e., Narrow Band IoT (NB-IoT) in 4G, massive machine type communications (mMTC), and ultra-reliable and low latency communications (URLLC) in 5G [12–14]. Further, the latest advancements in cloud computing and big data technologies enable fast and accurate analysis of huge volumes of structured and non-structured data, thus creating lots of business opportunities for the development of more sophisticated and intelligent IoT systems. The continuous progresses and widespread deployments of IoT technologies have been transforming many industrial sectors and commercial ecosystems. Now, IoT applications and services are becoming indispensable to our daily lives and business models, e.g., remote control of door locks, lights and appliances at homes and offices; real-time modeling of resource consumption patterns and streamline business processes in factories; constant surveillance for property security, public safety, and emergency management in cities.

To meet the fast-growing demands of various IoT applications and services for different businesses and customers, many leading ICT companies, such as Amazon, Google, Microsoft, Cisco, Huawei, Alibaba, MI and JD, have launched their own cloud-based IoT platforms for common data-centric services. However, these enterprise-level platforms are not designed for data sharing, nor service collaboration. General concerns of data security and customer privacy strictly prevent the attempts of connecting and integrating them for much bigger commercial benefits and global influences. Besides, it is even harder to overcome the existing barriers of vertical industries and realize cross-domain information exchanges for minimizing the redundancies and fragments at different but related IoT applications.

In the future, when AI technologies are widely adopted in most industrial sectors and application domains, new links will be established between those domain-specific island-like solutions. In most cases, they are not used to share original data, but only to exchange necessary knowledge that is purposely learned from separated/protected data sets for customized applications. To realize this ambitious vision, the key function and challenge for future IoT systems is innovative knowledge creation, which requires high-quality data, super-intelligent algorithms, and more computing resources everywhere. Centralized cloud computing alone cannot support this fundamental change, while dispersive fog computing technologies will fill the computational gap along the continuum from cloud to things. Therefore, future intelligent IoT systems will fully utilize the best available computing resources in their neighborhoods and in the clouds to calculate novel effective mechanisms and solutions, which are acceptable and executable across different enterprise platforms, industrial sectors, and application domains. In this way, those domain-specific IoT systems are not closed or isolated any more, they will become much more powerful and influential by working collaboratively, thus significantly saving global resources, improving overall performance, and maximizing potential benefits in all systems. We have no doubt that future IoT applications and services will be shifting from data-centric to knowledge-centric. Very soon, they will become better than human-designed ones, since they are taught by us and powered by accumulated data, sophisticated AI algorithms, endless computing resources, and fast evolutions. Eventually, they will help us not only search and identify useful information from massive raw data, but more importantly, discover and create new knowledge to expand our horizons and capabilities.

The rest of this chapter is organized as follows. Section 1.2 reviews some well-known IoT standards and technologies. Typical IoT applications are summarized in Sect. 1.3. New challenges and future directions of IoT systems are analyzed in Sect. 1.4. Finally, Sect. 1.5 concludes this chapter.

1.2 Standards and Technologies

The IoT refers to the interconnection and exchange of data among devices/sensors. Currently, with the explosive growth of the IoT technologies, an increasing number of practical applications can be found in many fields including security, asset tracking, agriculture, smart metering, smart cities, and smart homes. The short-range radio technologies (e.g., Radio-frequency identification (RFID), Near Field Communication (NFC), ZigBee) are widely used for building automation, automotive, and monitoring devices as well. For example, Wi-Fi based on the IEEE 802.11 standards was used in most office environments. However, the short-range radio technologies are not adapted for scenarios that require long range transmission.

Cellular networks are widespread and ubiquitous, covering 90% of the world's population, and other technologies like Wi-Fi do not have the same scale, requiring users to search for and connect to a local network. RF providers, wireless infrastructure companies, and carriers have made massive investments in cellular networks to provide secure, reliable service to as many customers as possible. By leveraging existing infrastructure and mature technology, cellular IoT can connect millions of IoT devices with little additional investment. Solutions based on cellular communications (e.g., 2G, 3G, and 4G) can provide larger coverage, but they consume excessive device energy.

IoT applications' requirements have driven the emergence of a new wireless communication technology: low power wide area network (LPWAN), as shown in Fig. 1.1 [15]. LPWAN is increasingly gaining popularity in industrial and research communities because of its low power, long range, and low cost communication characteristics. It provides long-range communication up to 10–40 km in rural zones and 1–5 km in urban zones. In addition, it is highly energy efficient (i.e., 10+ years

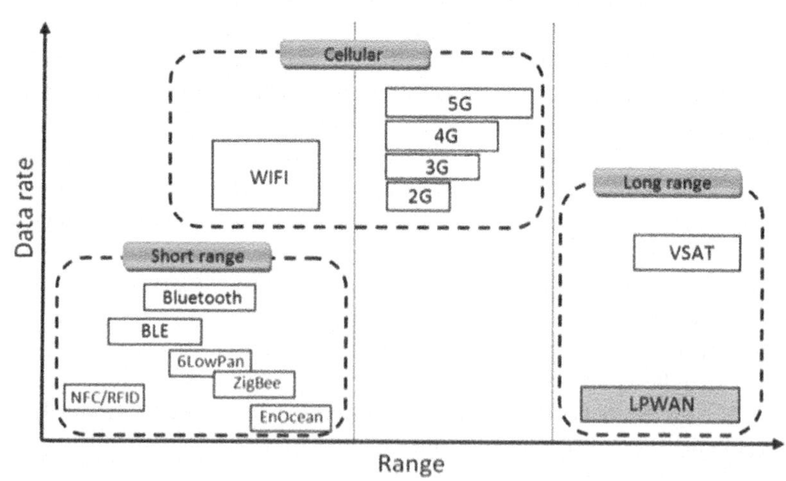

Fig. 1.1 Required data rate vs. range capacity of radio communication technologies

of battery lifetime) and inexpensive, with the low cost of a radio chip-set [15]. In summary, LPWAN is highly suitable for IoT applications that only need to transmit tiny amounts of data in long range. Many LPWAN technologies have arisen in the licensed as well as unlicensed frequency bandwidth. Among them, Sigfox, LoRa, and NB-IoT are today's leading emergent technologies that involve many technical differences [15].

Myriads of IoT connectivity solutions are available to support a wide range of IoT application with miscellaneous requirements. Therefore, in order to select an optimal technology for an application, various factors, such as power consumption, security issues, deployment cost, communication range, data rate, and latency, are required to be considered. A comparison of some typical IoT connecting solutions (i.e., RFID, NFC, ZigBee, LoRa, Sigfox, and NB-IoT) on pre-specified factor is given in Table 1.1. We will detail the connecting solutions in the next subsections. Further, considered a subset of the IoT, we will introduce Web of Things (WoT) which focuses on software standards and frameworks of IoT, such as REST, HTTP, and URIs, to create applications and services that combine and interact with a variety of network devices.

1.2.1 Radio Frequency Identification

The roots of radio frequency identification (RFID) date back to World War II. Germans, for instance, used an interesting maneuver in which their pilots rolled their planes as they return to base, so it would change the reflecting radio signal. This simple procedure alerted the ground radar crew of German planes returning and not allied aircraft. It can be considered one of the first passive ways to identify an object by means of a radio frequency signal, which was known as "identify friend or foe (IFF)." In the 1960s and 1970s, RFID systems were still embedded within a context of "secret technology." As an example, Los Alamos National Laboratory was asked by the Energy Department of United States of America to develop a system for tracking nuclear materials and control sensitive materials. In the 1990s, MIT's Auto-ID Center developed the global standard for RFID and other sensors, which described the IoT as a system where the Internet is connected to the physical world through ubiquitous sensors. In 2010s, the decreased cost of equipment and tags, increased performance to a reliability of 99.9%, and a stable international standard brought a major boost in the use of RFID systems.

There are two major initiatives regarding RFID standardization: International Standard Organization (ISO) and EPCglobal. ISO uses a cross-industry perspective with a generic approach. Meanwhile EPCglobal adopts a more application-specific approach. The most recognized outcome of EPCglobal is the establishment of Electronic Product Code (EPC), a unique code for item numbering, for identification of objects by using a similar approach as bar code numbering. ISO works closely with International Electro-technical Commission (IEC) responsible for general type of RFID standards covering issues of air interface, data content, conformance, and

Table 1.1 Comparison of IoT connecting technologies

Technology	RFID	NFC	Zigbee	LoRa	Sigfox	NB-IoT
Range	1–700 m	1–10 cm	75–100 m	2–15 km	3–50 km	10–15 km
Bandwidth	2–26 MHz	14 kHz	2 MHz	<500 kHz	100 Hz	180 kHz
Frequency band	125 kHz, 134.2 kHz, 13.56 MHz, 433 MHz, 860–960 MHz	13.56 MHz	868 MHz, 915 MHz, 2.4 GHz	868 MHz, 915 MHz, Sub 1 GHz	915–928 MHz, Sub 1 GHz band	700 MHz, 800 MHz, 900 MHz
Data rate	4–640 kbps	100–424 kbps	250 kbps	50 kbps with FSK	Less than 100 bps	200 kbps
Latency	1–10 ms	100 ms	≈15 ms	1–10 ms	10–30 ms	<10 s
Modulation scheme	OOK, FSK, PSK	ASK	O-QPSK	FSK	GFSK, DBPSK	BPSK, QPSK
Battery lifetime	Battery free	Battery free	1–5 years	>10 years	<10 years	>10 years
Cost	Low	Low	Low	Low	Low	Low
Application	Materials management, attendee tracking	Payment, ticketing	Smart home, healthcare	Air pollution monitoring, fire detection	Smart meter, pet tracking	Street lightning, agriculture
Mobility	Limited mobility	Limited mobility	Limited mobility	Yes	Yes	Nomadic
Advantages	High speed and convenience, and very low cost	Convenient to use	Highly reliable and scalable	Highly immune to interference, adaptive data rate	High reliability, low device complexity	The requirement of Gateway better range and coverage
Limitation	Brings security and privacy concerns	Limited data transfer rates	Short range, communication security issues	Supports limited size data packets, longer latency, not acknowledges all packets	Requires multiple transmissions, suffer from interference	No hand-off support, low interference immunity, lacks in acknowledging
Standard body	ISO/IEC	ISO/IEC	ZigBee Alliance	LoRa Alliance	Sigfox	3GPP

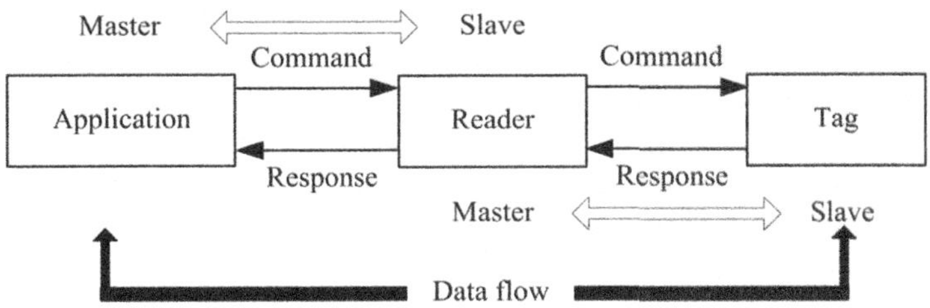

Fig. 1.2 The architecture of a RFID system

performance. ISO/IEC standards cover certain areas of technology. For instance, the ISO 18000 series standards are a series of standards that define air interface standards consisting of channel bandwidth, EIRP, modulation, data coding, and bit rate, ISO 15418 is a standard that defines data content. There are also many separate standards that were already developed for livestock tracking (ISO 11785, ISO 11784, and ISO 14223) [6, 16].

A typically RFID system is consisted of three main components: RFID tags, reader, application system [6, 17], as shown in Fig. 1.2. RFID uses electromagnetic fields to automatically identify and track tags attached to objects. The tags contain electronically stored information. The RFID tags are known as transponders (transmitter/responder), are attached to the objects to count or identify. Tags could be either active or passive. Active tags are those that have partly or fully battery powered, have the capability to communicate with other tags, and can initiate a dialogue of their own with the tag reader. Passive tags collect energy from a nearby RFID reader's interrogating radio waves. Active tags have a local power source (such as a battery) and may operate hundreds of meters from the RFID reader. Tags consist mainly of a coiled antenna and a microchip, with the main purpose of storing data. The reader is called as transceiver (transmitter/receiver) made up of a radio frequency interface (RFI) module and control unit. Its main functions are to activate the tags, structure the communication sequence with the tag, and transfer data between the application software and tags. The application system is known as data processing system, which can be an application or database, depending on the application. The application software initiates all readers and tags activities. RFID provides a quick, flexible, and reliable way for electronically detecting, tracking, and controlling a variety of items. RFID systems use radio transmissions to send energy to a RFID tag while the tag emits a unique identification code back to a data collection reader linked to an information management system. The data collected from the tag can then be sent either directly to a host computer, or stored in a portable reader and up-loaded later to the host computer.

RFID technology has many advantages. RFID tag and reader should not be in LOS to make the system work, and a RFID reader is capable of scanning multiple tags simultaneously. Unlike bar codes, RFID tags can store more information. Moreover, it follows instructions/commands of reader and provides location to the reader along with its ID. RFID technology is versatile in nature and hence smaller and larger RFID devices are available as per application. Tags can be read only as well as read/write unlike bar codes. However, RFID technology also has disadvantages. Active RFID is costly due to use of batteries. Privacy and security are concerns with the use of RFID on products as it can be easily tapped or intercepted. RFID devices need to be programmed which requires enough amount of time. The external electromagnetic interference can limit the RFID remote reading. The coverage range of passive tags is limited which is about 3 m.

RFID takes the market in many different areas including inventory management, personnel, and asset tracking, controlling access to restricted areas, supply chain management, counterfeit prevention. The addition of other sensors around AIDC—Automated Identification and Data Capture technologies such as infrared detectors, radiation, humidity, and others in RFID applications contributed to the development of IoT by extending it to reach intelligent services and providing local capabilities for actuation. For example, in manufacturing, RFID technology offers many applications in the automotive industry. The RFID-based anti-theft vehicle anti-theft device is a protection device installed in many vehicles. RFID also offers great promise for the assembly and manufacturing process of automobiles, especially for flexible and flexible production planning, spare parts, and inventory management. RFID technology not only helps automate the overall assembly process, but also significantly reduces costs and shrinkage, but also provides better service for automotive users, including more efficient replacement parts ordering and automatic generation of maintenance reminders. The benefits that RFID brings to the automotive industry, including production processes and end users, are visibility, traceability, flexibility, and enhanced security. In the supply chain, managers will be able to monitor the status of shipments like a crate filled with fruit. With sensors, RFID tags, and RFID readers, the manager sees the exact location of the crate inside the warehouse, the fruit's point of origin, days until expiration, and temperature in real time. A visible, transparent process improves efficiency, reduces waste, and allows traceability. If a shipment is determined to be unsuitable for consumption due to disease or other circumstances, the source or cause of the defection will quickly be discovered because of the great wealth of information available.

In summary, the adoption of RFID is spurring innovation and the development of the IoT, which are commonplace throughout households, offices, warehouses, parks, and many other places. Industry and government mandates are regulating RFID technologies leading to accepted standards across industries allowing for interoperability among devices. Additionally, the cost and size of devices continue decreasing which allows companies to embed smaller, common items with RFID chips and sensors. Although promising, RFID is not without its challenges, which arise from electromagnetic interference, security, and privacy issues. Communication between tags and readers is inherently susceptible to electromagnetic interference. Simultaneous transmissions in RFID lead to collisions as readers and tags typically operate on a same wireless channel. Therefore, efficient anti-collision protocols for identifying multi-tags simultaneously are of great importance for the development of large-scale RFID applications Due to its cost and resource constraint limitations, RFID system does not have a sufficient security and privacy support. Many researchers and scientists work to implement low cost security and privacy protocol to increase the applicability. Lots of lightweight solutions have been proposed for RFID, but they are still expensive and vulnerable to the security and do not fully resolve the security issues.

1.2.2 Near Field Communication

Near field communication (NFC) is a short-range wireless communication technology. NFC technology uses magnetic coupling to send and receive signals. When two NFC enabled devices are close enough (from touch to 10 cm), they create an electromagnetic field between them. That electromagnetic field allows the active NFC device to power up and communicate with the passive NFC device. The active NFC device then picks up on variations in signal levels specific to the passive device and reads those variations as a signal. A detector and decoder circuit in the active NFC device is then used to comprehend the passive NFC signal and extract the relevant information. NFC technology builds on RFID, which uses an ISO/IEC standard. NFC was approved as an ISO/IEC standard in 2003 and is standardized in ECMA-340 and ISO/IEC 18092. These standards specify the modulation schemes, coding, transfer speeds, and frame format of the RF interface of NFC devices, as well as initialization schemes and conditions required for data collision-control during initialization for both passive and active NFC modes. They also define the transport protocol, including protocol activation and data-exchange methods. NFC incorporates a variety of existing standards including ISO/IEC 14443 Type A and Type B.

The possible interaction styles among NFC devices provide three different operating modes as shown in Fig.1.3 [18]. Three types of NFC devices are involved in NFC communication: smartphones, NFC tags, and NFC readers. In reader/writer mode, an active NFC device can read, write, or change the stored data in a passive NFC tag. This mode is just like traditional RFID technology, where a terminal reads the data from the embedded chip on the smart card. The maximum possible data rate in this mode is 106 kbps. In peer-to-peer mode, two active devices can exchange small amount of data in between them. As both devices are battery powered they can establish radio link in between them. They set up bi-directional, half duplex channel having maximum data rate of 424 kbps. In card emulation mode, the active NFC devices work as a smart card based on ISO/IEC 14443 Type A and Type B. This model is compatible with preexisting smart-card industry.

NFC has advantages in IoT application. One of the notable benefits or advantages of NFC revolves around its simplicity and expansive applications. It is easier to set up and deploy than Bluetooth because it does not require pairing or manual configuration. The connection is automatic and takes a fraction of a second. It also uses less power than other types of wireless communication technologies. Another remarkable advantage of NFC is that it supports the widespread application of contactless payment systems. Several companies have implemented payment

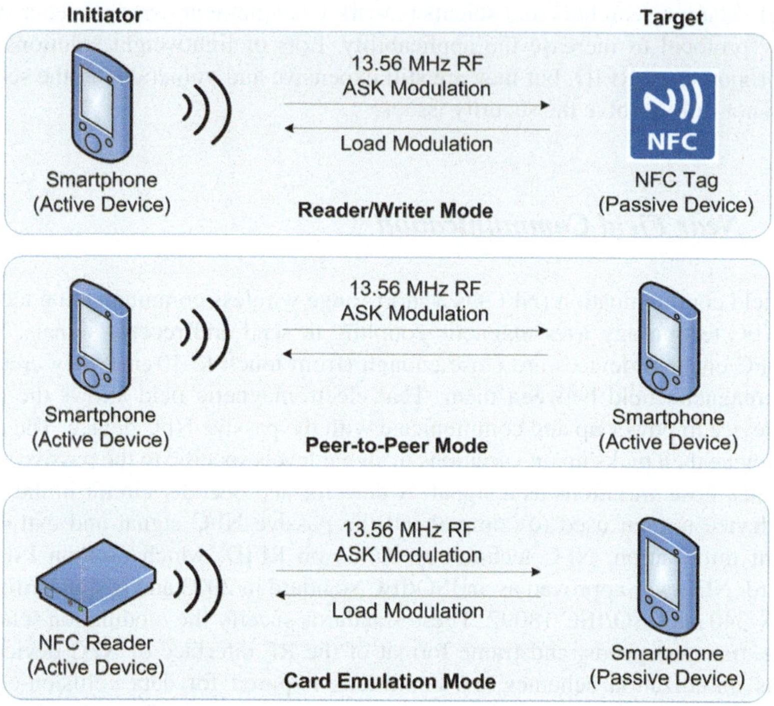

Fig. 1.3 NFC system architecture

transactions based on this technology. However, NFC also has disadvantages. It may prove to be much too expensive for companies to purchase and maintain related machines and other equipment in IoT applications. So small companies could find it difficult to sustain their existing turnover and enhance profits. Installing the hardware and software and hiring technicians to maintain the same could result in spiraling expenses for the concerned company. A critical limitation or disadvantage of near field communication is that it is not as effective and efficient as Bluetooth or Wi-Fi Direct when it comes to data transfer rates. NFC can only send and receive very small packets of data. Its maximum transfer rate is 424 kbps while Bluetooth 2.1 has a transfer rate of 2.1 Mbps. While NFC transactions are undoubtedly more secure than regular credit card payments, this technology is not completely free from risk. Rapid evolution in technology always comes with an equally powerful negative consequence. Mobile phone hacking is now rampant and attackers are coming out with newer methods to gain unauthorized access into users' personal, social security, and financial data stored therein. This makes the entire system vulnerable and insecure. The obvious lack of security could discourage both users and companies from warming up to this technology in the near future.

NFC offers great and varied promise in services such as payment, ticketing, gaming, crowd sourcing, voting, navigation, and many others. NFC technology enables the integration of services from a wide range of applications into one single smartphone. NFC technology is typically used for payments and marketing applications today. Payment using NFC integrated smart cards offers easier payment compared to conventional multiple step payment process. Top payment services like Visa and MasterCard are offering NFC embedded smart cards to customers. NFC integrated smart cards can be used for fast payments at grocery shops, parking tickets, adding shopping points, redeem coupons with just a single tap of the card. All the major banks around the globe offer smart cards with NFC chips integrated. NFC integrated system can be used in medicine and healthcare activities. NFC offers greater accuracy and convenience in prescribing medicine, easier check-in, payments, checking status of patients, tracking records by embedding NFC tags to patient's charts. NFC integrated devices can be easily paired and configured. Medical professionals can easily check schedules and access medical devices and equipment.

NFC is one major emerging technology of the last decade. Even though it remains a comparatively newborn technology, NFC has become an attractive research area for many researchers and practitioners due to its exponential growth and its promising applications and related services. From the technical point of view, some security issues in NFC technology are already solved and standardization is mostly provided as well. However, there are still unsolved security issues. For example, new protocols/mechanisms on off-line and on-line authentication of NFC tags should be studied. NFC specific alternative key exchange protocols should be proposed to prevent various attacks on RF communication.

1.2.3 ZigBee

ZigBee is a low cost, low data rate, and short distance wireless ad hoc networks standard which contains a set of communication protocols. ZigBee is developed by ZigBee Alliance based on IEEE 802.15.4 reference stack model and mainly operate in two different frequency bands: 2.4 GHz and 868/915 MHz. ZigBee is used to provide services such as small area monitor, security, discovery and profiling for industrial control, household automatic control, and other places which deployed sensor network-based applications.

The thought of ZigBee-style can be tracked to the end of 1990s which is proposed by IEEE 802.15 group. After that, IEEE 802.15.4 (TG) group devoted to the bottom standards. In 2001, ZigBee Alliance was founded by Honeywell and some other companies which aims at creating, evolving, and promoting standards for ZigBee. In 2004, ZigBee Alliance published the ZigBee 1.0 (a.k.a. ZigBee 2004) standards. Then, the ZigBee 2006 that revised in former version was published in 2006. In 2007, the alliance published ZigBee PRO standard which contains two sets of advanced commands. In 2009, a standard that is more flexible and has remote control capability named ZigBee RF4CE was announced. From 2009, ZigBee adopts IETF's IPv6 standard as the standard of smart energy and committed to forming a globally unified network.

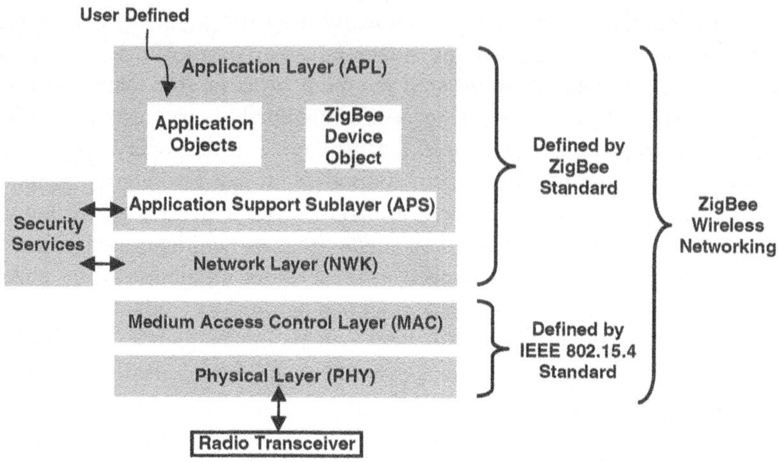

Fig. 1.4 ZigBee system architecture

In general, the system architecture of ZigBee is depicted by four layers, as shown in Fig. 1.4 [19]. As mentioned before, ZigBee is developed on the top of IEEE 802.15.4 standard. ZigBee standard is built on the two lower layers: the physical (PHY) layer and the medium access control (MAC) sublayer which were defined by IEEE 802.15.4, then ZigBee Alliance providing the network (NWK) layer and the framework for the application layer [9].

PHY IEEE 802.15.4 standard defines the PHY layer of ZigBee. The PHY layer is the lowest layer and defines the interface between PHY channel and MAC sublayer, it also provides data service and PHY layer management service. The PHY layer data service directly controls and communicates with the radio transceiver (transmitter and receiver). The PHY layer management service maintains the database which related to the PHY layer parameters. The protocol stipulates that the PHY layer operate in two separate frequency bands: 2.4 GHz and 868/915 MHz.

MAC IEEE 802.15.4 standard defines the MAC layer of ZigBee. The MAC layer provides the interfaces between the PHY layer and the NWK layer and controls access to the radio channel using a CSMA-CA mechanism. The MAC layer is responsible for establishment, maintenance and termination of wireless links, and synchronization between devices, and it transmits frames into the protocol system.

NWK The network layer provides interfaces between the MAC layer and the APL layer and is responsible for managing the network and routing. The NWK layer supports three topologies: star, tree, and mesh topologies. Managing the network includes network establishment, end devices discovery, join and departure, and these operations are controlled by the ZigBee coordinator. The ZigBee coordinator not only responsible for assigning network address for devices in network but also responsible for discovering and maintaining the path in network due to the terminal devices have no ability for routing. Routing is to choose the path to forward information to the target devices.

APL The APL layer is the highest protocol layer in ZigBee standard. The APL layer can be divided into three parts: application support sublayer (APS), application framework, and ZigBee device object (ZDO). The APL layer is responsible for mapping the variety of applications to ZigBee network, and it mainly includes service discovery, convergence of multiple data stream, and security.

Zigbee has many advantages, and the main characteristics of ZigBee are low data rate, low power consumption, low complexity, high security and support for a variety of network topologies. However, there still exits some disadvantages: The cost for the end devices is difficult to reduce at present, that makes it is not cheap when deploy a large number of end devices. The communication distance is only about 75–100 m, that is because the communication frequency mainly operates in 2.4 GHz. Signals at the frequency are difficult to penetrate through blocks, which seriously affect the communication distance.

ZigBee is used to provide services such as small area monitor, security, discovery and profiling, and so on for industrial control, household automatic control, and other places which deployed sensor network-based applications. For example, ZigBee can be used in the building energy consumption monitoring system due to low cost, devices sparsity, low energy consumption, and self-organized characteristics. The end devices which equipped with different sensors are used to monitor the temperature, humidity and voltage, and so on. The end devices also can collect the data from water meters, gas meters, and electricity meters. These data which is gathered from variety of end devices will be sent to the upper computer, then the policy will be made by the special system to achieve the goal which includes energy consumption monitoring, temperature control, and energy-saving operation management.

In summary, ZigBee provides a short distance, low complexity, low energy consumption, low data rate, and low cost technology for wireless network and it effectively compensates for the vacancies in the low cost, low power, and low rate wireless communication market. ZigBee also has many facets that can be improved, and we believe that the ZigBee technology never stops its step and there will be more and more ZigBee-based applications emerge in our life.

1.2.4 LoRa

LoRa (short for long range) is an emerging technology in the current market, which operates in a non-licensed band below 1 GHz for long-range communication link operation. LoRaWAN defines the communication protocol and the system architecture, while LoRa defines the physical layer. LoRa Technology offers compelling features for IoT applications including long range, low power consumption, and secure data transmission. The technology can be utilized by public, private, or hybrid networks and provides greater range than cellular networks. The bandwidth has been established to ensure data rates from 0.3 kbps up to 50 kbps, which is not much compared with IEEE 802.11 but enough for the majority of applications in automation and data collection field, and also ensuring maximization of the battery life in case of mobile or autonomous terminals. The concept is really affordable for IoT applications, especially because of the reduce cost of implementation in long range conditions.

LoRa was invented by a startup in France called Cycleo whose employees are veterans of big semiconductor companies who wanted to build a long range low power communication device. They filed a patent in 2008 titled "Fractional-N Synthesized Chirp Generator" and another in 2013 titled "Low power Long range transmitter." Later this company was acquired by another French company named Semtech that's into manufacturing of analogue and mixed-signal semiconductors. Semtech's LoRa Technology has amassed over 600 known uses cases for smart cities, smart homes and buildings, smart agriculture, smart metering, smart supply chain and logistics, and more. 97 million devices are connected to LoRa networks in 100 countries and the number of the devices is still growing. While Semtech provides the radio chips featuring LoRa Technology, the LoRa Alliance, a non-profit association and the fastest growing technology alliance, drives the standardization and global harmonization of LoRaWAN, which is a media access control (MAC) protocol for LoRa [10]. To fully define the LoRaWAN protocol, and to ensure interoperability among devices and networks, the LoRa Alliance develops and maintains documents to define the technical implementation, including MAC layer commands, frame content, classes, data rates, security, and flexible network frequency management, and so on.

A LoRa server architecture consists of end nodes, LoRa gateway, LoRa network server, and LoRa application server. LoRa end nodes are the sensors or application where sensing and control take place. These nodes are often placed remotely. The nodes are the devices sending data to the LoRa network server. These devices could be, for example, sensors measuring air quality, temperature, humidity, location, and so on. The LoRa gateways are different from cellular communication where mobile devices are associated with the serving base stations. The gateways are receiving data from the devices and typically run an implementation of the packet-forwarder software. This software is responsible for the interface with the LoRa hardware on the gateway. The LoRa server component provides the LoRaWAN network server component, responsible for managing the state of the network. It has knowledge of devices active on the network and is able to handle join-requests when devices want to join the network. When data is received by multiple gateways, LoRa network server will de-duplicate this data and forward it once to the LoRaWAN application server. When an application server needs to send data back to a device, LoRa network server will keep these items in queue, until it is able to send to one of the gateways. LoRa application server provides an API which can be used for integration or when implementing your own application server. The LoRa application server component implements a LoRaWAN application server

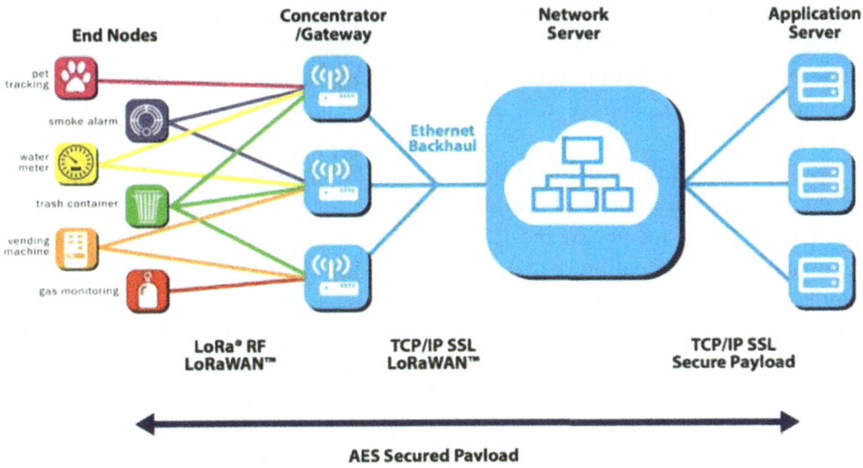

Fig. 1.5 LoRa network architecture

compatible with the LoRa server component. It provides a web-interface and APIs for management of users, organizations, applications, gateways, and devices (Fig. 1.5).

Everyday municipal operations are made more efficient with LoRa Technology's long range, low power, secure, and GPS-free geolocation features. By connecting city services such as lighting, parking, waste removal, and more, cities can optimize the use of utilities and personnel to save time and money. LoRa Technology and smart city IoT networking can offer street light solutions that increase energy efficiency and reduce city operating costs. LoRa solutions are easy to implement into existing infrastructure and allow smart monitoring of the grid over a LoRaWAN network, LoRaWAN street light controller is LoRaWAN-alliance network compatible street light control system for street lights. The system provides a unique identity to every light, allows independent control of street light on calendar and timer basis, and allows instant manual control of lights from software control system. LoRa Technology's low power qualities and ability to penetrate dense building materials make it an ideal platform for IoT-connected smart home and building devices. In addition, the long range capabilities make it possible for LoRa-enabled sensors to track assets that stray from home. Sensors in smart home and building applications can detect danger, optimize utility usage and more to improve the safety and convenience of everyday living. LoRa-enabled products can include thermostats, sprinkler controllers, door locks, leakage monitors, and smoke alarms. These devices connect to a building's network and allow consistent, remote monitoring to better conserve energy and predict when maintenance is necessary, saving property managers money. LoRa Technology has the capacity to function in high density environments, such as in large enterprise buildings or campuses, and can handle thousands of unique messages per day.

LoRa has advantages in IoT applications. LoRa uses industrial, scientific, and medical (ISM) bands 868/915 MHz which is globally available. It has very wide coverage range about 5 km in urban areas and 15 km in suburban areas. It consumes less power and hence battery will last for longer duration. Single LoRa Gateway device is designed to take care of 1000s of end devices or nodes. It is easy to deploy due to its simple architecture. LoRa uses adaptive data rate technique to vary output data rate output of end devices. This helps in maximizing battery life as well as overall capacity of the LoRaWAN network. The physical layer uses a spread spectrum modulation technique derived from chirp spread spectrum (CSS) technology. This delivers orthogonal transmissions at different data rates. Moreover, it provides processing gain and hence transmitter output power can be reduced with same RF link budget and hence will increase battery life. However, LoRa also has disadvantages. LoRaWAN network size is limited based on parameter called as duty cycle. It is defined as percentage of time during which the channel can be occupied. This parameter arises from the regulation as key limiting factor for traffic served in the LoRaWAN network. LoRaWAN supports limited size data packets and has longer latency. So it is not ideal candidate to be used for real-time applications requiring lower latency and bounded jitter requirements.

In summary, the use of LoRa is strongly influenced by the characteristics of the environment in which the technology is implemented. The type of IoT application to be used needs to pay attention to these things. The use of LoRa in open areas such as rural areas has advantages in power usage, wide range, and flexibility in network development. However, for applications that require specific requirements such as the amount of data exchanged per specific time period then, the network configuration needs to be considered primarily for indoor implementations. The LoRa network cannot transmit large amounts of data for a wider range of territories. LoRa Technology is influenced by obstacles such as tall buildings and trees, which leads to an increase in packet loss levels in the zone. The use of GPS in the LoRa module has not been reliable, especially for real-time position tracking software applications. The limitations identified in LoRa Technology become an opportunity for future research.

1.2.5 Sigfox

Sigfox is a French global network operator founded in 2009 that builds wireless networks to connect low power objects such as IoT sensors and smartwatches, which need to be continuously on and emitting small amounts of data [11, 20]. Sigfox wireless technology is based on LTN (Low Throughput Network). It is wide area network based technology which supports low data rate communication over larger distances. It is used for M2M and IoT applications which transmits only few bytes per day. By employing ultra-narrow band in sub-GHz spectrum, Sigfox efficiently uses the frequency band and has very low noise levels, leading to very low power consumption, high receiver sensitivity, and low cost antenna design. Sigfox had developed a simple communications protocol, running in the license-free ISM bands

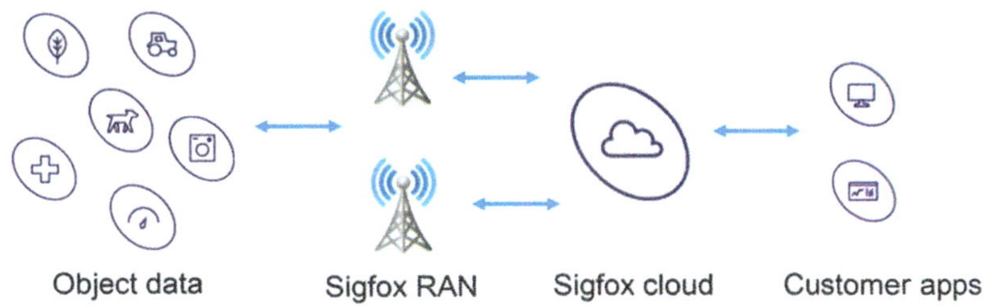

Fig. 1.6 Sigfox network architecture

at 868 and 915 MHz. It uses very low cost, standard chips, has a usable range of 5–10 km and a battery life which can support years of low data rate transmission. At first glance, it looked like the answer to an IoT maiden's dream. The only problem they had was how to deploy it. Unlike the roll-out of cellular connectivity, where you could start with the areas of greatest use—typically capital cities, the IoT customer base is much more diverse. Cities come into it, particularly if they are trying to be smart, but there are plenty of applications in agriculture and transport which need much wider coverage. Building their own network would have taken too long and cost too much, so they persuaded mobile operators to partner with them and install Sigfox's gateways on their existing towers, providing a fairly rapid coverage. As of October 2018, the Sigfox IoT network has covered a total of 4.2 million square kilometers in a total of 50 countries and is on track to reach 60 countries by the end of 2018.

Figure 1.6 depicts simple Sigfox network architecture [11, 20]. The Sigfox network consists of objects (end user devices), Sigfox gateway or base stations, Sigfox cloud and application servers. Sigfox objects are connected with gateway using star topology. There is direct secure point-to-point link between Sigfox gateways and Sigfox cloud. The cloud interfaces with servers using different protocols such as SNMP, MQTT, HTTP, IPv6, etc., as per end applications. Sigfox offers a software based communications solution, where all the network and computing complexity is managed in the cloud, rather than on the devices. All that together, devices connected through the Sigfox network only use the network when they are actually required to transmit data, in this procedure much of the power consumption is reduced.

Sigfox has advantages in IoT applications. Sigfox has designed its technology and network to meet the requirements of mass IoT applications; long device battery life-cycle, low device cost, low connectivity fee, high network capacity, and long range. A device is not attached to a specific base station. Its broadcasted messages are received by any base station in the range, which is 3 on average, and there is no need for message acknowledgement. UNB intrinsic ruggedness coupled with spatial diversity of the base stations offer great anti-jamming capabilities. UNB is extremely robust in an environment with spread spectrum signals. Low bit rate and simple radio modulation enable a 163.3 dB budget link for long-range communications. SIGFOX has tailored a lightweight protocol to handle small messages. Less data to send means less energy consumption, hence longer battery life. With its simple approach to connectivity, Sigfox provides extremely price-competitive connectivity subscriptions and even more importantly, enables extremely simple and cost-efficient silicon modules. Sigfox is compatible with Bluetooth, GPS 2G/3G/4G, and Wi-Fi. By combining other connectivity solutions with Sigfox, business cases and user experience can be drastically improved. However, Sigfox also has disadvantages. The narrow band spectrum emitted by single Sigfox end device causes strong interference and collision to nearby existing wideband system. More such Sigfox devices will further enhance the interference. Sigfox supports one-way communication without acknowledgement. This necessitates multiple transmissions if server does not receive data without errors. Due to the multiple transmissions, power consumption will increase which depends on number of retransmissions. Due to low data rate support, it cannot be used for high data rate applications.

Nowadays, pallets tracking to determine the location and goods condition are highly desirable in logistics. In this application, the most requirements are low cost sensors and long battery lifetime for asset tracking and status monitoring, in this case Sigfox is a good solution. Logistics companies can have their own network so they have a guaranteed coverage in their facilities. Low cost Sigfox devices could be easily deployed on vehicles. Sigfox public base stations can be then used when vehicles are outside the facilities or when goods arrive at customer locations. In the retail and hospitality industries, keeping guests satisfied and customers engaged is your number one priority. Now, the IoT is here to help. Sigfox-enabled IoT solutions for retail and hospitality change the game by keeping you connected to all aspects of your retail location, hotel, restaurant or Powered by Sigfox's network dedicated exclusively to the IoT, the latest IoT solutions improve upon earlier versions of connectivity technology to provide a cost-efficient, user-friendly experience.

In summary, Sigfox is rolling out the first global IoT network to listen to billions of objects broadcasting data, without the need to establish and maintain network connections. This unique approach uses a compact and optimized wireless protocol to reduce signaling overhead, based on which Sigfox nodes are not attached to the network. However, in order to support a myriad of devices in IoT, the interference between Sigfox devices and nearby existing wideband system should be mitigated in further research. Further, new technologies and mechanisms should be investigated to reduce retransmissions. Sigfox system works well in fixed location. There are issues such as interference and frequency inaccuracies in the mobility environments, which also presents challenges to further research and application.

1.2.6 NB-IoT

NB-IoT is a LPWAN technology based on narrowband radio technology. The technology provides improved indoor coverage, support of massive number of low throughput devices, low delay sensitivity, ultra-low device cost, low device power consumption, and optimized network architecture. The technology can be deployed by utilizing resource blocks within a normal LTE carrier, or in the unused

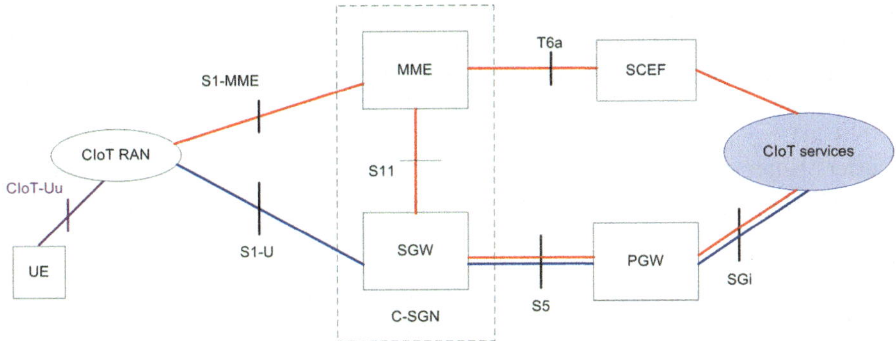

Fig. 1.7 NB-IoT network architecture

resource blocks within a LTE carrier's guard-band, or standalone for deployments in dedicated spectrum. NB-IoT is standardized by the 3rd Generation Partnership Project (3GPP). Its specification was published at Release 13 of 3GPP on June 2016 [21]. In December 2016, Vodafone and Huawei integrated NB-IoT into the Spanish Vodafone network and send the first message conforming to the NB-IoT standard to a device installed in a water meter. NB-IoT now has received strong support from Huawei, Ericsson, and Qualcomm. The objectives of NB-IoT are to ensure a device cost below 5 USD, uplink latency below 10 s, up to 40 connected devices per household, a device with 164-dB coupling loss, and a 10-year battery life can be reached if the user equipment transmits 200 bytes of data a day on average. NB-IoT has entirely an extensive ecosystem that is available globally. This is primarily due to its support from more than 30 of the world's largest and top class operators. These operators have global communication coverage that serves above 3.4 billion customers and geographically serve over 90% of the IoT market [22].

NB-IoT network architecture is shown in Fig. 1.7. In order to send data to an application, two optimizations for the cellular Internet of Things (CIoT) in the evolved packet system (EPS) were defined, the User Plane CIoT EPS optimization and the Control Plane CIoT EPS optimization [23]. Both optimizations may be used but are not limited to NB-IoT devices. On the Control Plane CIoT EPS optimization, data are transferred from the eNB (CIoT RAN) to the MME. From there, they may either be transferred via the serving gateway (SGW) to the packet data network gateway (PGW), or to the service capability exposure function (SCEF) which however is only possible for non-IP data packets. From these nodes they are finally forwarded to the application server (CIoT Services). DL data is transmitted over the same paths in the reverse direction. In this solution, there is no data radio bearer set up, data packets are sent on the signaling radio bearer instead. Consequently, this solution is most appropriate for the transmission of infrequent and small data packets. The SCEF is a new node designed especially for machine type data. It is used for delivery of non-IP data over control plane and provides an abstract interface for the network services (authentication and authorization, discovery, and access network capabilities). With the User Plane CIoT EPS optimization, data is transferred in the same way as the conventional data traffic, i.e., over radio bearers via the SGW and the PGW to the application server. Thus it creates some overhead on building up the connection; however, it facilitates a sequence of data packets to be sent. This path supports both, IP and non-IP data delivery.

NB-IoT has advantages in IoT applications. As it uses mobile wireless network it offers better scalability, quality of service, and security compare to unlicensed LPWAN such as LoRa/Sigfox. It offers long battery life due to low power consumption or current consumption. NB-IoT also offers extended coverage compare to GSM/GPRS systems and co-exists with other legacy cellular systems such as GSM/GPRS/LTE. The NB-IoT compliant devices can be deployed/scheduled within any legacy LTE network. This helps them share capacity as well as other cell resources with the other wireless connected devices. NB-IoT modules are expected to be available at moderate costs. It offers better penetration of structures and better data rates compare to unlicensed band based standards (e.g., LoRaWAN and Sigfox). However, NB-IoT also has disadvantages. It offers lower data rate (about 250 Kbps download and 20 Kbps upload) compare to LTE. The bandwidth is about 200 KHz. Hence it is ideal to use NB-IoT for stationary devices. NB-IoT devices need to connect to an operator network via licensed spectrum. Network and tower handoffs will be a problem, so NB-IoT is best suited for primarily static assets, like meters and sensors in a fixed location, rather than roaming assets.

The strong growth in the NB-IoT market has motivated many analyst firms to create forecasts showing the expected numbers of connections as well as the revenue potential. Generally, the global IoT market is expected to be worth trillions of dollars by 2020. The NB-IoT market is a subset of this, and it is important for operators to understand the revenue potential in the countries they operate in. Humans and their pets share a good bond, unfortunately many users often face issues regarding lost or stolen pets. Pet tracking use case is one application that helps the user to keep track of its pets' activities and most importantly location at all times. A small lightweight device placed around the neck of the pet embedded with an NB-IOT chip-set helps to send tracking information to its user's device. The NB-IOT device collects and sends location information leveraging GPS and location based services and this can be done either periodically or in real time based on the users' preferences. The user can then receive the information with a tracking route that is already integrated with the map. Furthermore, this device is embedded with several forms of alarms that can alert the user when the device battery is running low [24]. Security has always been a very important aspect of human living, people at all times want to be guaranteed of home safety [24]. Alarms and event detection will help to rapidly inform that user about a detected home intrusion. NB-IoT system will not only offer intelligent protection from intrusion but will also offer intelligence for detected events that can lead to a fire outbreak like a sudden increase in home temperature or smoke. Alarms and events detectors will make use of sensors placed devices in ideal locations in the home that constantly communicates with the NB-IoT network, this use case will make use of a very low data throughput and battery life of the devices will be ultra-critical.

In summary, NB-IoT is a promising technology for IoT applications. It can connect low power IoT devices that are placed in weak coverage environments such as apartment basements. This is done by allowing devices to repeat signal transmissions while operating at very low power. Compared to LoRa and Sigfox, NB-IoT relies on the existing cellular infrastructure instead of new ones, thus the investments on a utility-dedicated communication infrastructure and the time required for deployment of applications are reduced. However, it is difficult to implement firmware-over-the-air (FOTA) or file transfers. Some of the design specifications for NB-IoT make it such that sending larger amounts of data down to a device is hard.

1.2.7 Web of Things

The emergence for the concept of Web of Things (WoT) benefits from the development and application of World Wide Web (WWW). In parallel the development of WWW, more and more sensors and devices which have computing and communication capabilities are deployed in some special areas or around persons. Connect these devices with Internet for data acquisition and action control is the basic thought of IoT. However, varieties of devices with different functions may adopt incompatible protocols, so that the heterogeneous network is isolated. Then, Web of Things has been proposed to avoid this situation. The WoT benefits from what made the web so successful and applies it to the embedded devices in order to unify the IoT standards [25]. The goal of WoT is connecting the things to the web using the protocols such as HTTP and HTML that adopted in the web and makes the applications have common interfaces which are consistent with developing web [26].

The start of WoT can be tracked to 2000, and a project which uses URLs to address the physical things and uses HTTP protocol to interact with things appeared in 2002. The application of IoT that invoke web was proposed by Dominique Guinard and Vlad Trifa in 2007, they started the WoT online community and published the WoT manifesto. In 2011, Guinard and Trifa founded EVRYTHNG which is a company use WoT to serve industry needs, and other startups also started the same time. In 2014, the World Wide Web Consortium (W3C) organized the workshop on the WoT and lead to the creation of the WoT Interest Group and the submission of the Web Tings Model. In the same time, other company started projects about the WoT, such as Google and Siemens.

The system architecture of WoT is described in four layers (These four layers' architecture is defined by functionality): the accessibility layer, findability layer, sharing layer, and composition layer as shown in Fig. 1.8 [25].

Accessibility Layer The accessibility layer is the lowest layer in the system architecture. Varieties of things offer web APIs so that they can be accessed to the Internet through the access layer. The layer is responsible for turning things into programmable web things. The operation orders are like this: physical things expose their services through RESTful APIs (REST refers to a set of

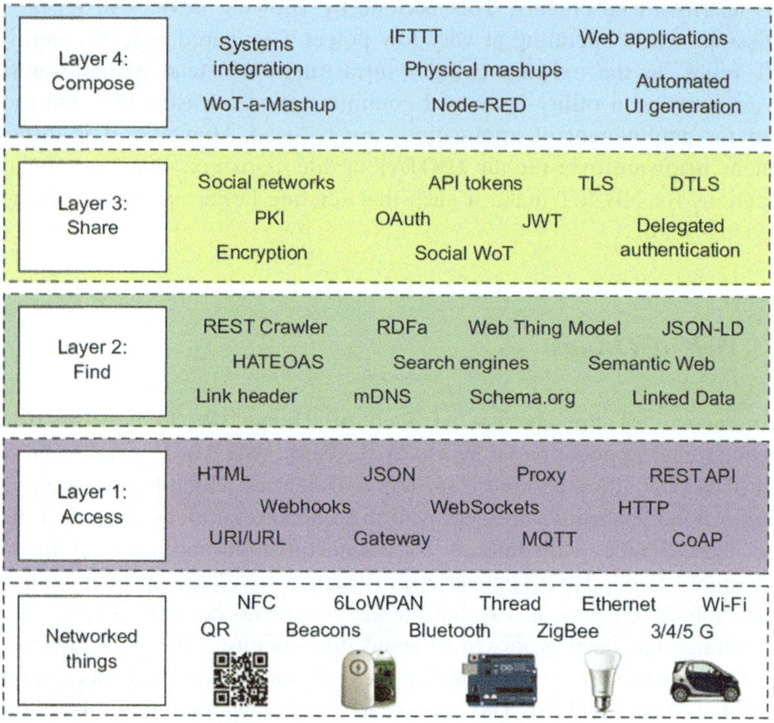

Fig. 1.8 WoB network architecture

architectural constraints and principles for web services. Web services that satisfy these constraints and principles are RESTful web services). If any real-world things can provide RESTful APIs over HTTP, then they will get URLs and integrate these things into the existing World Wide Web (WWW).

Findability Layer The real-world things have become seamlessly part of web after the accessibility layer. However, other applications or users do not know what the services have been provided by these web things. So the findability layer is responsible for describing things and their services that enabled by semantic annotations and make them findable and useable. This layer makes the things can be searched through the web indexes and search engines by other applications.

Sharing Layer The real-world things can be searched and shared by humans and applications through several ways on the Internet. Then, how can users obtain and use the things' data efficiently and safety? The sharing layer takes the responsibility for establishing the mechanisms to ensure the appropriate data can be shared and used safety. While the HTTP protocol already provides the mechanisms for securely sharing data in WWW, there needs more perfect authentication mechanisms for

WoT due to the new characteristics. A promising mechanism which uses fine-grained sharing has been proposed on the top of RESTful APIs [25].

Composition Layer The composition layer is the highest layer in the system architecture. The responsibility of this layer is simplifying the process of creating new applications which are made from things and web services. The mashups process can use some web tool-kits (such as JavaScript SDKs), dashboards that with programmable widgets and mashup tools (such as Node-RED). These tools make it convenient for users and applications.

WoT is a promising settlement that can break the heterogenic isolation dilemma in IoT and it is a specification of IoT by integrating the things into web architecture. All the resources and services are formed into web service APIs for humans and applications that will simplify the process of applications development. However, shortcomings are inevitable, such as data security and local privacy. WoT as a unified web that makes it more vulnerable than IoT networks. WoT is not the ultimate solution for connecting all things and perceiving the world.

The range that WoT will be applied is wide, such as industrial control, household device control, smart city and environmental monitor. WoT makes that easier for varieties of devices and network interconnection. Many applications have been developed based on WoT architecture. For example, in a smart home case, many types of sensors and cameras are deployed in a home for monitoring some parameters, such as brightness, temperature, and shoot video. WoT will adopt its four layers' architecture to integrate the sensors into the WWW, so that the hardware of the sensors can be presented at web. Meanwhile, the resources will be provided through web service APIs form. Then, users can do their mashups more easily.

In the past few years, the devices which have computing and communication capacities have experience an explosive growth, and the growth will continue in the future. WoT was expected to solve the all things interconnection problem, and some progress has been made. Nevertheless, there still exits many problems to settle in the present, such as devices accessibility, sharing mechanisms designation, and data authority and privacy. Although the WoT is faced with some problem, it is still a promising architecture in the future, and we believe the new applications of WoT will spring up.

1.3 Typical Applications

1.3.1 Environmental Monitoring

Environmental monitoring is an important application area of the IoT. The automatic and intelligent characteristics of the IoT are very suitable for monitoring environmental information. Generally speaking, the structure of the environmental monitoring based on IoT includes the following parts: The perception layer: The main function of this layer is to obtain environmental monitoring information,

Fig. 1.9 Sensor (left) and gateway (right) at the bottom of a birdhouse

such as temperature, humidity, and illumination, through sensor nodes and other sensing devices. Figure 1.9 (see Fig. 12 in [27]) shows the application of IoT in birdhouse monitoring. Because environmental monitoring needs to be perceived in a wide geographical range and contains a large amount of information, the devices in this layer need to be formed into an autonomous network through wireless sensor network technology, extract useful information by means of collaborative work, and realize resource sharing and communication through access devices with other devices in the Internet. The accessing layer: The main function of this layer is to transfer information from the sensing layer to the Internet through the wireless communication network (such as wired Internet network, ZigBee, LPWAN, WLAN network, GSM network, TDSCDMA network), satellite network, and other infrastructure. The network layer: The main function of this layer is to integrate the information resources within the network into a large intelligent network that can be interconnected, and establish an efficient, reliable, and trusted infrastructure platform for the upper layer of service management and large-scale environmental monitoring applications. The service management layer: The main function of this layer is to conduct real-time management and control of the massive information obtained by environmental monitoring within the network through a large central computing platform (such as high-performance parallel computing platform), and provide a good user interface for the upper application. The application layer: The main function of this layer is to integrate the functions of the bottom layer of the system and build the practical application of the industry oriented to environmental monitoring, such as real-time monitoring of ecological environment and natural disasters, trend prediction, early warning and emergency linkage, etc. Through the above parts, environmental monitoring based on IoT can realize collaborative perception of environmental information, conduct situation analysis, and predict the development trend.

1.3.2 Infrastructure Health Monitoring

The deployment of IoT is important for infrastructure health monitoring. As an engineering feat, China's massive South-to-North Water Diversion Project is a stunner. The massive Internet-of-Things (IoT) network that has been quietly overseeing the middle route is impressive in its own right. More than 100,000 individual sensors stud the 1400 km waterway, which connects the Danjiangkou reservoir to Beijing and Tianjin. For the last year, it has been scanning the canal for structural damage, tracking water quality and flow rates, and watching for intruders, whether humans or animals. There are many challenges in the South-to-North Water Diversion Project. The water traversed regions prone to earthquakes, making infrastructure vulnerable to damage. The water's flow would need to be controlled so that none of it would go to waste. Its quality would also need to be checked periodically to ensure no pollutants or toxins made their way into city drinking water supplies. In some places, local villagers scaled the fence to fish or swim in the water. That created safety risks. After careful analysis, these challenges are divided into three broad categories—infrastructure, water, and security—and, after some discussions, settled on more than 130 different kinds of Internet-connected sensors to install along the canal. Infrastructure sensors measuring stress, strain, vibration, displacement, earth pressure, and water seepage were embedded in the ground adjacent to the canal and in the concrete banks and bridges as well as the 50 dams built to control the water's flow. Probes that measure water quality and flow rate were attached to the steel support columns that hold up the bridges. Video cameras were spaced every 500 m along the entire structure, as shown in Fig. 1.10. The smart gateway was developed to receive data continuously from local sensors and then transmit it to a cloud server using whatever signal was available at the moment. That could include fiber, Ethernet, 2G, 3G, 4G, Wi-Fi, or ZigBee. The smart gateway periodically transmits data to the nearest server, which may be any one of 47 regional branch servers in counties along the canal. Under normal circumstances, the transmissions occur at intervals of 5–30 min, or once a day, depending on location and water resources in the area. If some special event happens such as an earthquake or a chemical spill, the data will be sent immediately and continuously to the cloud. From there the data is stored or forwarded to any of the five administrative servers in provincial cities between the Danjiangkou reservoir and Beijing, to ultimately reach the main server center in Beijing. A Web platform and interface are designed. They allow people working at the server stations to read the data and respond to any alerts via a website, thus enabling the central management team in Beijing to always learn the latest developments at remote sites and make the right decisions in real time. Meanwhile, because the network is isolated from the World Wide Web, there is less of a risk that the data will be hacked by outsiders.

Fig. 1.10 Internet-connected sensors installed along the canal include video cameras (left), as well as intrusion detection (top right) and water level sensors (bottom right)

1.3.3 Public Safety Surveillance

The public safety surveillance based on IoT with the characteristic of wide coverage of public security monitoring, multiple monitoring indicators, high continuity requirements, unsuitable environment for manual monitoring, and close correlation between perceived information content and people's lives employs the technology of IoT especially the technology of sensor network to construct an information system engineering composed of perception layer, network layer and application layer, which mainly includes the monitoring to ensure the safety of all kinds of production scenes, the monitoring of producer safety, the monitoring of the safety of specific items, the monitoring of densely populated places, the monitoring of important equipment and facilities, and the information collection of scenes, personnel, and items during the emergency treatment of accidents. Public security is the cornerstone of national security and social stability. In order to effectively withstand all kinds of man-made or natural disasters, countries will strengthen public security measures as the focus of government work. The IoT for public safety monitoring provides a new way to solve the problems faced about public safety surveillance at present. The establishment of a complete public safety surveillance based on IoT will provide effective prevention mechanism for existing safety problems such as bridge tunnel collapse, hazardous material leakage, etc. The nationwide public safety surveillance based on IoT enables the timely, powerful and transparent resolution of major safety incidents. Therefore, the public safety surveillance based on IoT should be given priority by the whole society. Figure 1.11 describes the network architecture of public safety surveillance based on IoT, which

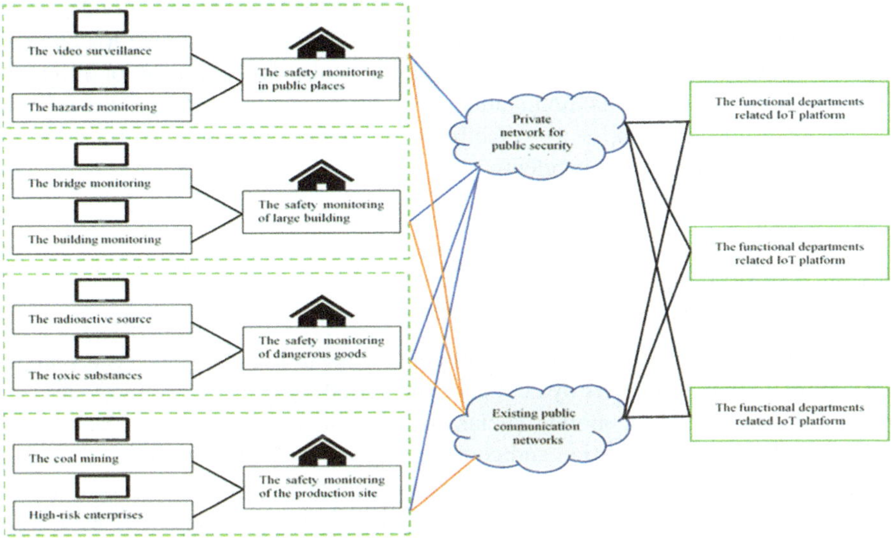

Fig. 1.11 The network architecture of public safety surveillance

is similar to the whole architecture of IoT and consists of three parts: perception layer, network layer, and application layer. However, due to the particularity of the application scene of public safety surveillance based on IoT, it has some technical characteristics that other IoT applications do not have, which are summarized as follows:

In the perception layer, the types of perceived information are diverse, and the real-time requirement is high. The perception of most information (such as the safety of bridge buildings, monitoring of dangerous goods, etc.) requires high accuracy and is difficult to detect by manual means. Because of the high uncertainty of the information type of potential security hazards, a large number of different types of sensors should be deployed in densely staffed or high-risk production sites for a long time. Higher requirements are put forward for the networking strategy of perception layer, energy management, transmission efficiency, QoS, sensor coding, address, frequency, and electromagnetic interference. These problems are also the key to the mature application of public safety surveillance based on IoT. Because the information perceived by public safety surveillance based on IoT involves the national key industries and the daily life of the people, once the information is leaked or improperly used, it may endanger national security, social stability, and people's privacy. Therefore, it is necessary for the information content of public safety surveillance based on IoT to be transmitted through the private network or 4G mobile network after taking security precautions to ensure the security, authenticity, and integrity of the information. It is necessary to establish a proprietary platform for public safety surveillance based on IoT with different levels in view of the massive data information and the serious harm that security hidden danger may bring. The service platform not only has a strong ability of information processing and integration, but also timely links relevant functional departments to deal with emergencies in case of public safety emergencies, so as to minimize losses and impacts. In addition, the interconnection of public security IoT platforms of different levels is conducive to the maximum allocation of resources according to the hazard level of security incidents, so as to facilitate the timely, effective, and transparent resolution of public security incidents.

1.3.4 Smart City

It is expected that 70% of the world six billion peoplewill live in cities and surrounding regions's population, over by 2050. So, cities need to be smart, if only to survive as platforms that enable economic, social, and environmental well-being. Smart city is the one that uses information and communications technologies (ICTs) to make the city services and monitoring more aware, interactive, and efficient. Smartness of a city is driven and enabled technologically by the emergent IoT, a radical evolution of the current Internet into a ubiquitous network of interconnected objects that not only harvests information from the environments (sensing) and interacts with the physical world (actuation/command/control), but also uses existing Internet standards to provide services for information transfer, analytics, and applications. As illustrated in Fig. 1.12, the smart cities have six characteristics: smart economy, smart people, smart governance, smart mobility, smart environment, and smart living. It is implied that the capability to achieve measurement, understanding, and visualization of multiple urban environment parameters is a key criterion in developing a smart city.

A number of specific application domains have also been identified that could utilize smart city IoT infrastructure to service operations in health services (noise, air, and water quality), strategic planning (mobility), sustainability (energy usage),

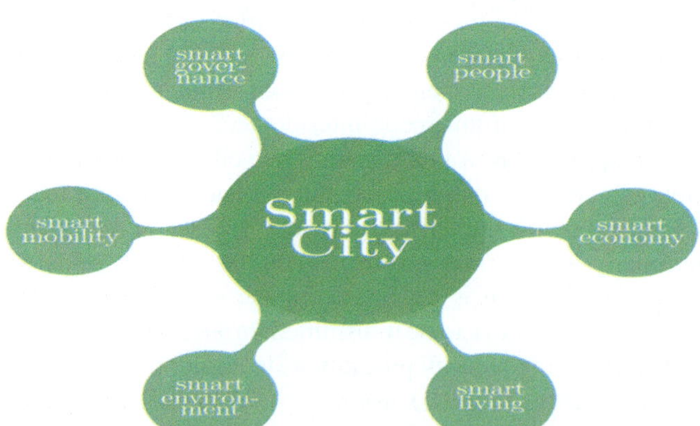

Fig. 1.12 The characteristics of smart cities

tourism (visitor services and tourist activity), business and international (city usage and access), and city safety. The end goal of smart city IoT platform is to have plug-and-play smart objects that can be deployed in any environment with an interoperable backbone allowing them to blend with other smart objects around them. In order to realize this goal, there are many technological hurdles including architecture, energy efficiency, security and privacy, QoS, cloud computing, data analytics, and GIS-based interpretation. Standardization of frequency bands and protocols plays an important role in accomplishing this goal. Several projects and activities detailed above are addressing these critical challenges in the next decade, a clearer picture regarding the usefulness of IoT in making the smart city will emerge. Due to the scale of activities, participation of large companies and the government will play a pivotal role in the success of this emerging technology.

1.3.5 Smart Manufacturing

With the increasing scale of industry, there have many issues and restrictions in the process of development, especially in some particular industry, such as the manufacturing of nuclear fusion device. Fusion device is generally the piecework and the manufacturer prefers to use traditional manufacturing process than to develop intelligent manufacturing process. The traditional manufacturing technology restricts the development of fusion device as a result of low automation and backward production management. There has a major impact on manufacturing and upgrading of fusion device through improving the management level. Product quality tests are also an important link to improve production efficiency. Therefore, improving the management of the manufacture, assembly, and testing level is the key to increase the quality of product. The application of intelligence will greatly improve the efficiency of manufacturing system. The intelligent manufacturing system is widely used to indicate the design innovations. With the fast development of information technology and intensification of trend in economic globalization, highly competitive business environment forces the managers to continuously make the best decisions in the shortest possible time. The application of IoT has penetrated into various industries. Based on IoT technology, the intelligent manufacturing system can greatly facilitate the production efficiency, improve the management level, and expand production scale. Intelligent manufacturing system has become the development trend of manufacturing technology. Unquestionably, IoT will bring high impact to manufacturing field. The interconnecting smart objects on the global network can be realized though Internet technologies and the development of corresponding supporting technologies. The evolution of IoT also brings a lot of challenges for many fields. Manufacture, as a very important application field nowadays, has been involved into IoT. Based on IoT, intelligent manufacturing system allows the combination of the virtual word and physical word anytime and anywhere. The purpose is to improve the context which the users needed and adopting the resources based on the requirements. The architecture of IoT for

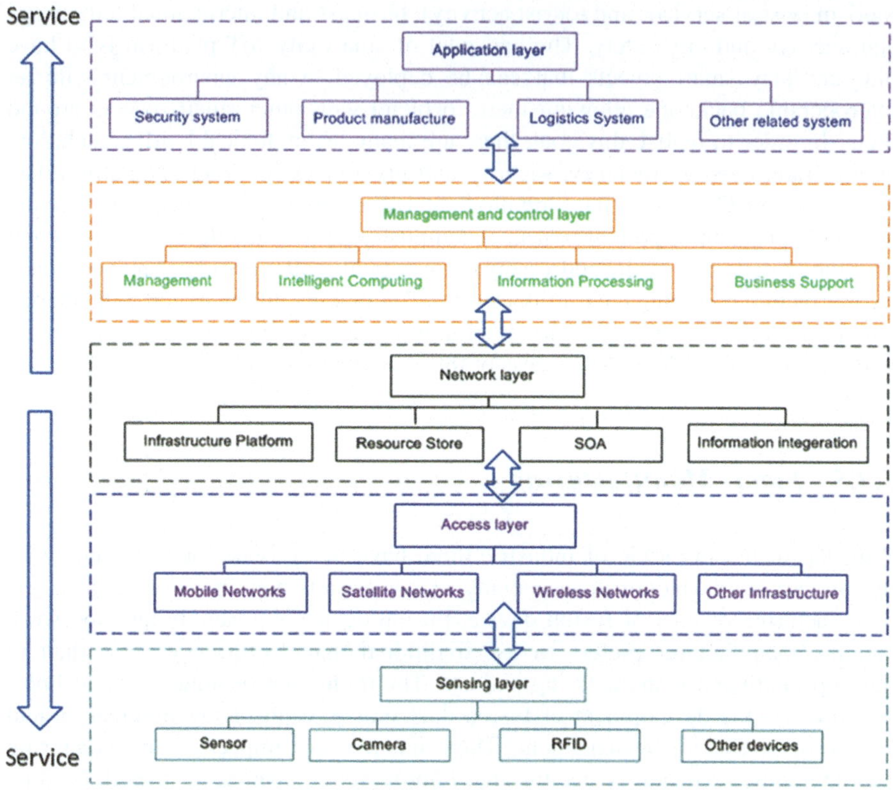

Fig. 1.13 The architecture of IoT for manufacturing field

manufacturing field has been proposed, as shown in Fig. 1.13 (see Fig. 1 in [28]). The architecture of IoT for manufacturing field includes application layer, management and control layer, network layer, access layer, and sensing layer where each layer contains a functional aspect. The application layer of this architecture consists of product manufacture, logistic system, and other system related manufacturing system. Its main function is to integrate the bottom systems and establish the various applications. The management and control layers consist of various platforms in order to provide a well users' environment to support the application of upper layer, such as management platform, information processing platform, intelligent computing platform, and so on. The network layer is to integrate the information resources from access layer into a larger intelligence network though the Internet platform and build a reliable and efficient infrastructural platform for upper layers. The access layer is responsible for transferring information to network layer by various networks, such as wireless networks, mobile networks, satellite networks, wireless LANs, and other infrastructure. The sensing layer is to acquire information by various sensors, RFID, and other identify intelligently. Then, it will share the information in the network. The service is throughout the entire architecture of IoT for manufacturing field.

1.3.6 Intelligent Transportation System

With the acceleration of urbanization, motorization, the pace of modernization, the urban population increase, growing faster vehicles, urban traffic congestion and clogging growing, urban transport system overwhelmed, the consequent environmental noise, air pollution, energy waste and other factors plaguing today's transportation in major cities have become a serious problem around the world now in industrial countries and developing countries. So, with the advance of information technology and globalization, the traditional means of transportation technology can no longer meet the requirements of economic and social development. The intelligent transportation is the inevitable choice for urban transport development, and is a revolution in urban transport undertakings. Things appear to intelligent transportation industry breakthrough brings rare opportunities to bring new horizons for the development of intelligent transportation, smart transportation to provide a wider space for development, and therefore modern urban transport calling for "Internet of things." "The new generation of intelligent transportation" developments provide important technical support for the realization of real-time, efficient, accurate, safe, energy-saving intelligent transport objectives and provide technical support for the IoT technology, networking technology will bring a new upgrade for intelligent urban traffic.

Roadside infrastructure, vehicles, and end users will form a network based on Intelligent Transportation System on IoT as shown in Fig. 1.14 (see Fig. 13.1 in [29]). The functions of the network in the transportation system are multifaceted:

1. Traffic departments can obtain and release traffic information in time through the network to effectively manage traffic infrastructure and road traffic.

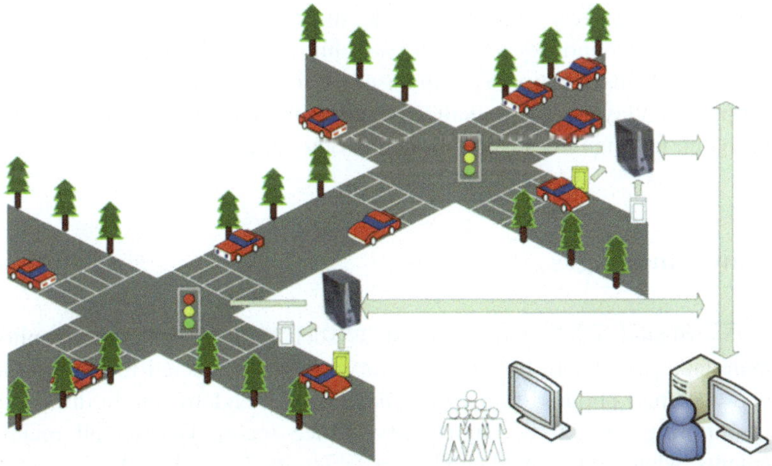

Fig. 1.14 Intelligent transportation system scenario overview

2. The traffic infrastructure can automatically adjust intelligently according to the traffic condition.
3. Traffic participants can obtain more comprehensive traffic information, travel advice, and road services and obtain efficient and safe traffic services.
4. In terms of energy conservation and environmental protection, the more efficient operation of the entire transportation network will improve energy efficiency and reduce environmental pollution.

However, due to the particularity of the application scene of Intelligent Transportation System based on IoT, it has some technical characteristics that other IoT applications do not have, which are summarized as follows:

1. The data acquisition of the sensing layer is mainly used to collect the state and data of objects on the road, including the position information and running state of various vehicles, the state of various intelligent mobile terminals on the road, and the real-time data of some roadside sensors. Intelligent Transportation System based on IoT data collection involves sensor, RFID, wireless communication, multimedia information collection, and other technologies. Sensor network networking and collaborative information processing technology realize short distance transmission of data acquired by sensor, RFID and other data acquisition technologies, self-organizing network, and collaborative processing of data information by multiple sensors.
2. The network layer realizes a wider range of interconnection functions and can transfer the information collected by the perceptive layer to the application layer of the Intelligent Transportation System based on IoT without any barrier, with high reliability and security. Meanwhile, the Intelligent Transportation System based on IoT data and instructions can also be transferred to the corresponding devices of the perceptive layer in an efficient, real-time, and safe manner. It requires the integration of sensor network, mobile communication technology, and Internet technology. With the development and application of LTE mobile communication technology and Internet technology, it can better meet the needs of Intelligent Transportation System based on IoT data transmission.
3. The application layer mainly includes intelligent navigation, intelligent parking, safe driving, intelligent transportation, and other application systems.

1.4 New Challenges

IoT has contributed a lot in different areas such as environmental monitoring, smart manufacturing where the entire factory can be connected to the Internet be controlled via smart phone. IoT has also played a tremendous role in infrastructure health monitoring and public safety surveillance sector. Despite all mentioned achievements, what are the main challenges facing the IoT? [30–34]:

1. Many of devices in IoT networks will require their own batteries. The lifetime of the devices is limited by the capacity of the batteries. Will energy scavenging and enormously low power circuits eliminate the need for batteries? Can the devices harvest energy from the energy resource in surrounding environment, such as wind, solar, thermal energy, and kinetic energy, to enable the realization of self-sustainable communication? Research must be conducted in order to develop systems that are able to harvest energy from the environment and optimize the energy consumption of a IoT system. This means that it is not reasonable to waste the energy by unnecessary data transmissions, protocol overheads, and so on.
2. As many things are connected to Internet it is necessary to have an adequate architecture that permits easy connectivity, control, communications, and useful applications. This has created a complex interoperability and integration challenge to realize large-scale heterogeneous IoT ecosystems. The communications between these heterogeneous devices need to be adaptive to allow dynamic interconnectivity and support decentralized nature. Trillions of objects will be connected by the Internet. Many things or set of things must be disjoint and protected from other devices. Therefore, it is important to protect, authenticate, authorize, and name these diverse devices.
3. In IoT there will exist a vast amount of raw data being continuously collected. It will be necessary to develop techniques to convert raw data into usable knowledge. For example, this can be more helpful in medical stream by monitoring the person heart rate, pulse, blood pressure and that raw data should be converted into usable knowledge by giving precautions to person or doctor like medical streams. It can be implemented in many fields like industrial, home appliances. On the other hand, for some applications, such as target tracking, ocean monitoring, etc., we will require multi-source data sensing, which will result in data redundancy. Therefore, we will face the challenge of data fusion inevitability. If the data of the same type of sensor is directly integrated from the data layer, that is, data layer fusion. If the sensor is not homogeneous and the data is heterogeneous, then feature fusion or decision fusion is used to mask the heterogeneity of the data itself. If multiple sensors are observing the same source of information, the resulting data must be correlated, so correlation operations are required.
4. For IoT, there is no single wireless technology that optimizes operation costs, bandwidth, range, power, and architecture for all applications and requirements. Even if two devices share the same wireless, they may not share the same communication protocol. Will a Panasonic refrigerator talk to a LG washing machine, a Lenovo computer, an Apple iPhone, an OSRAM smart LED, and a Samsung TV? Probably it is impossible for a while. However, the smart gateway is a better selection. It has two functions, one is the protocol conversion, and the other is wireless routing. The information can wirelessly interact between the different smart terminals by using the smart gateway. In the smart home, there is a wireless routing function to realize information aggregation and controlling of the smart terminal.
5. With IoT gradually becoming a reality, the huge amount of data collected from billions of devices could possibly paralyze big data and information technology

(IT) companies. Presently, many companies have already had some difficulties to organizing and analyzing their existing data. They are also worried about the number and type of devices to be connected to the Internet. In fact, current IT infrastructures that are stretched to their limits today may come crashing down as we approach the IoT/IoE era. In order to overcome these problems, big data and IT companies need to use data-reduction methods such as model building, sampling, and aggregation. They could also use visualization techniques or visual analytics to analyze the results.

6. Security not only provides an essential pillar for Internet but also introduces a major challenge for the IoT. As the time goes the trend of IoT inflates from millions of devices to tens of billions. As increasing the number of connected devices, the chance to exploit security vulnerabilities is also increase, like in cheap or low standard designed devices, and due to incomplete data streams the chances of data theft is increased. Many IoT arrangements will also include collections of similar or adjacent similar devices. This homogeneity expands the potential impact of any single security weakness by the total number of devices that all have the same features. In IoT, the security challenge is mainly embodied in the following aspects:

   - Many IoT systems have both hardware and software vulnerability that remain unpatched. If a hacker exploits those vulnerabilities, there will be zero-day attacks. That could be a disaster for the entire organization, and it will be hard to mitigate those attacks because the manufacturers were not aware of vulnerabilities.
   - IoT is exposed to larger attack surface. Because devices are connected each other, many attacks are possible not to one device but to the entire network.
   - Consumers have low knowledge of IoT; people enjoy IoT, but few understand how it works and they do not pay attention to the security issues.

   The fundamental problem that is pervasive computing in the Internet today that must be solved is dealing with security attacks. Security attacks are problematic for the IoT because of the minimal capacity of things be used, the physical accessibility sensors, actuators and objects, and the openness of the system, including the fact that most devices will communicate wirelessly. Backdoor is the most concern in IoT security which can be caused by vendors while updates of things happen. Identifying and naming of the object is also an important thing in IoT. And use of wireless sensor networks plays a crucial role in IoT which may leads to security issues.

7. Authenticity, trustworthiness, and confidentiality are important for IoT. Currently, the data networks are still delicate. The cloud storage operation is still in the emerging stage. Transmit the data to a cloud service for processing, sometimes includes a third party. The gathering of this information leaks legal and regulatory challenges facing data protection and privacy law. In order to realize the opportunities of the IoT, some new strategies will be required to address privacy issues, and innovate technologies and services will be developed through a broad range of expectations. Privacy is the most concern in IoT, the data

which storing in cloud using big data should not be seen by any other person. To solve these problems privacy policies for each system should be specified. Once specified either the individual IoT applications or the IoT infrastructure must enforce privacy.

8. In IoT, many applications work on the basics of sensing, automation, and computation platform. In these deployments, it is common for devices to know their locations, have synchronized clocks, know their neighbor devices when cooperating and have coherent set of parameter settings such as consistency, sleep, awake schedules, and appropriate power levels for communication. How to achieve cross-platform information integration and how to enhance security and privacy will be the challenges that the future IoT needs to face. Looking at the development of IoT, we can see that data collection is the foundation, information extraction is the key point, knowledge creation is the core, and cross-border integration is the key. As an extension to cloud computing, fog computing enables service provisioning along the continuum from the cloud to things for reducing latency and bandwidth demands, and for empowering end users in their vicinity [5]. Specifically, fog computing can pool resources anywhere along this continuum and can deploy its service anywhere in this range, including in the cloud, at the edge or on the things. It is foreseeable that based on fog computing, mass sensory terminal deployment, and mass IoT information storage, the development of IoT is bound to shift from informatization and intelligence to knowledge-based. In order to promote the development of intelligent IoT, a flexible and ubiquitous software/hardware platform has become an urgent need to achieve cross-sectoral and cross-platform information integration.

1.5 Conclusion

This chapter has reviewed some well-known IoT technologies and related standards, including RFID, NFC, ZigBee, LoRa, Sigfox, NB-IoT, and WoT. Their system architectures and technical advantages are briefly discussed. Then, some typical IoT applications are introduced. When more and more IoT systems are deployed for different industrial sectors, it is very challenging to overcome the vertical barriers and mitigate the fragmentation problem across multiple application domains. It is also very difficult to guarantee system security and customer privacy while connecting and integrating several enterprise-level IoT platforms with heterogeneous data structures.

At the same time, we regard these challenges as new research and development opportunities, especially when advanced communication technologies, fog and edge computing resources, and sophisticated AI algorithms are available in the neighborhoods of different application environments. As the focus of 5G standardization and commercialization is shifting to two IoT scenarios, cloud, fog and edge computing technologies will jointly make future IoT systems and applications more and more intelligent and innovative.

References

1. International Telecommunication Union (ITU) (2005) ITU Internet reports 2005: the Internet of things-executive summary
2. International Telecommunication Union (2016) Overview of the Internet of things. http://handle.itu.int/11.1002/1000/11559
3. Yang Y, Huang F, Ge X, Zhang X, Gu X, Guizani M, Chen HH (2006) Double sense multiple access for wireless ad hoc networks. In: Proceedings of the 3rd international conference on quality of service in heterogeneous wired/wireless networks (QShine 06). ACM Press, New York
4. Yang Y, Wu HH, Chen HH (2006) SHORT: shortest hop routing tree for wireless sensor networks. In: 2006 IEEE international conference on communications. IEEE, Piscataway
5. Zhao C, Zhang W, Yang Y, Yao S (2015) Treelet-based clustered compressive data aggregation for wireless sensor networks. IEEE Trans Veh Technol 64(9):4257–4267
6. Jia X, Feng Q, Fan T, Lei Q (2012) RFID technology and its applications in internet of things (IoT). In: 2012 2nd international conference on consumer electronics, communications and networks (CECNet). IEEE, Piscataway
7. Li L, Xiaoguang H, Ke C, Ketai H (2011) The applications of Wi-Fi-based wireless sensor network in internet of things and smart grid. In: 2011 6th IEEE conference on industrial electronics and applications. IEEE, Piscataway
8. Vedat C, Ok K, Busra O (2012) Near field communication: from theory to practice. Istanbuh NFC Lab-Istanbul, ISIK University, Turkey, p 82–94
9. Alliance Z (2012) Zigbee specification. http://www.zigbee.org/wp-content/uploads/2014/11/docs-05-3474-20-0csg-zigbee-specification.pdf
10. Augustin A, Yi J, Clausen T, Townsley W (2016) A study of LoRa: long range & low power networks for the internet of things. Sensors 16(9):1466
11. Sigfox (2018) Sigfox device ARIB mode white paper. https://support.sigfox.com/docs/sigfox-device-arib-mode-white-paper
12. Narrowband Internet of Things (NB-IoT) (2017) Technical report for BS and UE radio transmission and reception (release 13). https://www.3gpp.org/dynareport/36802.htm
13. 3GPP (2016) Cellular system support for ultra-low complexity and low throughput Internet of Things (CIoT). https://portal.3gpp.org/desktopmodules/Specifications/SpecificationDetails.aspx?specificationId=2719
14. Yang Y, Xu J, Shi G, Wang CX (2018) 5G wireless systems. Springer, Berlin
15. Mekki K, Bajic E, Chaxel F, Meyer F (2019) A comparative study of LPWAN technologies for large-scale IoT deployment. ICT Express 5(1):1–7
16. Adhiarna N, Rho JJ (2009) Standardization and global adoption of radio frequency identification (RFID): strategic issues for developing countries. In: 2009 fourth international conference on computer sciences and convergence information technology. IEEE, Piscataway
17. Finkenzeller K (2010) RFID handbook: fundamentals and applications in contactless smart cards, radio frequency identification and near-field communication. Wiley, London
18. Coskun V, Ozdenizci B, Ok K (2015) The survey on near field communication. Sensors 15(6):13348–13405
19. Farahani S (2011) ZigBee wireless networks and transceivers. Newnes, New South Wales
20. Sigfox (2018) Sigfox device ETSI mode white paper. https://support.sigfox.com/docs/sigfox-device-etsi-mode-white-paper
21. 3GPP (2016) Standardization of NB-IOT completed. https://www.3gpp.org/news-events/1785-nb_iot_complete
22. Huawei (2016) NB-IoT white paper. http://carrier.huawei.com/en/technical-topics/wireless-network/NB-IoT/NB-IoT-White-Paper
23. Schlienz J, Raddino D (2016) Narrowband internet of things whitepaper
24. Huawei (2016) NB-IOT enabling new business opportunities whitepaper. https://e.huawei.com/en/material/his/iot/6ba6590551ed4ad8b7bbe3c751fe8ea4

25. Guinard D, Trifa V (2016) Building the web of things: with examples in node.js and raspberry pi. Manning, Shelter Island
26. Ma J, Jin H, Yang, LT, Tsai JJ-P (eds) (2006) Ubiquitous intelligence and computing. Springer, Berlin
27. Lazarescu MT (2013) Design of a WSN platform for long-term environmental monitoring for IoT applications. IEEE J Emerging Sel Top Circuits Syst 3(1):45–54
28. Zhang R, Liu X, Liu L, Zhang X (2014) Research on an intelligent manufacturing system for tokamak machine. J Fusion Energy 33(6):648–652
29. Muthuramalingam S, Bharathi A, Gayathri N, Sathiyaraj R, Balamurugan B (2019) IoT based intelligent transportation system (IoT-ITS) for global perspective: a case study. In: Internet of things and big data analytics for smart generation. Springer, Berlin, p 279–300
30. KrishnaKanth Gupta and Sapna Shukla (2016) Internet of things: security challenges for next generation networks. In: 2016 International conference on innovation and challenges in cyber security (ICICCS-INBUSH). IEEE, Piscataway
31. Nzabahimana JP (2018) Analysis of security and privacy challenges in internet of things. In: 2018 IEEE 9th international conference on dependable systems, services and technologies (DESSERT). IEEE, Piscataway
32. Yadav EP, Mittal EA, Yadav H (2018) IoT: challenges and issues in Indian perspective. In: 2018 3rd international conference on internet of things: smart innovation and usages (IoT-SIU). IEEE, Piscataway
33. Yeo KS, Chian MC, Ng TCW, Tuan DA (2014) Internet of things: trends, challenges and applications. In: 2014 international symposium on integrated circuits (ISIC). IEEE, Piscataway
34. Farhan L, Shukur ST, Alissa AE, Alrweg M, Raza U, Kharel R (2017) A survey on the challenges and opportunities of the internet of things (IoT). In: 2017 eleventh international conference on sensing technology (ICST). IEEE, Piscataway

Chapter 2
Fog Computing Architecture and Technologies

2.1 Introduction

IoT is now emerging to support tens of billions of resource-limited devices such as smartphones around us connected to the networks [1]. It is driving a digital transformation in all aspects of our lives and businesses. Future IoT devices could, in particular, be widely deployed for tasks such as environmental monitoring, city management, and medicine and healthcare, requiring data processing, information extraction, and real-time decision making. The growing number of connected devices is creating data at an exponential rate, and increasingly more applications have strict delay requirements. The current computing paradigm which heavily relies on the cloud platform in addition to the host itself will not be able to keep up with this trend and meet the following challenges [2]:

1. Low Latency Requirement: Many applications in IoT have rigorous service delay requirements, especially for the industrial use cases and the Internet-of-Vehicles. Particularly, the smart manufacturing systems necessitate end-to-end latencies in the order of millisecond. The vehicle-to-vehicle communications and the drone flight controls typically require latencies below a few tens of milliseconds. All of these scenarios will not be handled adequately with only the remote clouds.
2. Limited Link Bandwidth: With the growing number of wirelessly connected IoT devices, the link bandwidth is also getting more and more congested. It becomes obvious that we should not try to convey all the information, e.g., the collected data at the sensors, to the remote clouds for further processing. On the one hand, the wireless spectrum is limited and we do not have enough bandwidth to send all the data to the cloud. On the other hand, researches indicate that most of the data generated by the end-points can be processed locally without jeopardizing the overall system performance.
3. Limited Computing Power: The IoT devices typically have limited computing power due to the cost and energy constraints. The current solution is to offload

Y. Yang et al., *Fog-Enabled Intelligent IoT Systems*,
https://doi.org/10.1007/978-3-030-23185-9_2

the computation burden to the remote clouds. However, this will inevitably incur additional communication and processing delays to the offloaded tasks. In addition to offloading to the remote clouds, we should try to offload the computing task to the neighboring edge nodes as well as to exploit the available computing power in the network effectively.

Fog computing provides the missing link in the cloud-to-thing continuum. Fog computing selectively moves compute, storage, communication, control, and decision making closer to the network edge where data is being generated in order to solve the limitations in current infrastructure to enable mission-critical, data-dense use cases. By definition [3], fog computing is a horizontal, system-level architecture that distributes computing, storage, control, and networking functions closer to the users along a cloud-to-thing continuum.

According to the definition, fog computing is a scenario where a huge number of heterogeneous, ubiquitous, and decentralized devices communicate and potentially cooperate among them and with the network to perform storage and processing tasks without the intervention of third parties [4]. Therefore, fog infrastructure will not only be at the network perimeter but also span along the cloud-to-things continuum, including in the cloud, at the edge, or on the things, and to also pool these distributed resources to support applications [2]. According to this trend in fog infrastructure, compute, storage, and networking services are the building blocks of the cloud and the fog that extends it. In this way, the processing of latency-sensitive applications can take place at the edge of the network, while resource hungry applications with delay-tolerant and computational intensive can take place in the cloud. Hence, with the ability to enable processing to take place at the network edge as specific locations near the end devices by the so-called fog nodes (FNs), the fog can provide advantage of low latency. Furthermore, ubiquitous FNs such as base stations (BSs), access points (APs), and routers positioned at the network edge offer densely distributed points for gathering data and task computation. With the advantages of fog computing, it is widely acknowledged that cloud computing is not viable for most of the IoT applications and fog could be used as an alternative [5].

Evolving from cloud computing, fog computing and cloud computing are not alternative but interdependent. However, they are different from each other. At first, due to the centralized feature, cloud computing is used for global tasks, so the resource can be optimized in a global view, whereas fog computing is used for scheduling and managing the local tasks. Also, due to close integration with the intelligence enabled front-end devices, fog computing is capable of efficiently enhancing the overall system performance [6]. In addition, fog computing extends a substantial amount of data storage, computing, communication, and networking of cloud computing near to the end devices. Although there are many cloud services that have already been applied for commercial use, the unreliable wireless connections, e.g., deep fading, often lead to packet loss and intolerable wide area network (WAN) delay between mobile devices and clouds [7]. Therefore, fog computing is a viable solution to the resource-constraint devices. Last but not least, although the fog computing achieves significant performance improvement in end-to-end latency, the reliability of the applications is still better in cloud computing.

Another similar concept we need to distinguish is edge computing. Cloudlet is one of the first edge computing concepts, which is cloudlet-based cyber foraging and provides computing and resource storage closer to the edge [7]. Then, based on the constrained resource, it allows mobile devices to offload computation in one or more virtual machines (VMs). Thanks to the VM technology, cloudlets provide cloud service to the mobile users instead of providing Internet connectivity as in Wi-Fi [8], thereby guaranteeing real-time application services. Using cloudlets, the resource-poor mobile users can find their preferable cloudlets to offload their intensive computations. Moreover, the cloudlets can act as a full cloud on the edge in a stand-alone environment, since cloudlets can be supported without the intervention of the cloud. Despite the benefits of cloudlets, it is hard to fulfill the QoS of the mobile devices as cloudlets are not an integral part of a mobile network [9]. However, fog computing tightly linked to the existence of a cloud, it cannot operate in a stand-alone mode.

To overcome the drawbacks of cloudlet and provide cloud computing capabilities at the edge of mobile network and within the radio access network (RAN), the industry specification group (ISG) within ETSI integrated edge computing into the mobile network architecture in 2014, which can be outlined as mobile edge computing (MEC) [10]. In March 2017, the ETSI has expanded the scope of MEC and after that replaced the term “mobile” by “multi-access.” The edges of non-mobile networks are also being considered in multi-access edge computing (MEC) [9]. Similarly, a key difference between MEC and fog computing is that MEC functions only in stand-alone mode, how it could interact with a distant cloud is still an open issue.

In conclusion, it should be noted that cloudlets, MEC, and fog computing aim at computing at edge, however, there is a significant difference between these technologies that need to be pointed. First, MEC is mainly driven by an industry consortium. According to the aim of OpenFog consortium, it focuses on the development of fog architecture and standardization details. Thus, fog and cloudlets are driven by research and development. Another difference is that cloudlets rely only on VM for virtualization, while both fog and MEC can consider other virtualization technologies other than VMs [11]. In addition, cloudlet mainly focuses on mobile offloading, whereas MEC aims to handle more applications from either mobile or non-mobile edge networks that are better provisioned. However, through the optimization of offloading decisions and the involved resource allocation to satisfy a substantial amount of latency-sensitive applications for resource-constraint end devices, fog computing overlaps between edge and cloud [11]. Finally, note that MEC works only in stand-alone mode until today. With regard to the cloudlets, although there is a lack of detailed literature review about how to interact between cloud and cloudlets, the cloudlets can function in either stand-alone mode or cloud mode by connecting to the cloud. Different from the above approaches, fog is designed as an extension to the cloud, which needs the support of the cloud for the task computation that cannot be computed in the resource-constraint fog layer.

Due to the elastic architectures and enabling technologies, fog computing has great potential in many application scenarios. But as any new technology in infancy, there also exist many challenges for practical fog deployments. In this chapter, we will discuss fog computing architectures and technologies. More specifically, we will first introduce fog computing architectures in industry and academia, termed OpenFog reference architecture and multi-tier computing network, respectively. Then, we will discuss the enabling technologies for fog deployments, including networking technologies, computing technologies, storage technologies, virtualization technologies, and artificial intelligence (AI). Finally, we will describe some detailed use cases of fog computing and look into how they would benefit from fog computing and what challenges would exist as well.

2.2 Fog Computing Architecture

2.2.1 Reference Architecture

OpenFog reference architecture for fog computing was first published by OpenFog Consortium (OPFWP001.0216, OPFRA001.020817) in 2016, 2017 [3, 12], and further adopted by IEEE standard (IEEE Standard 1934–2018) in 2018 [13]. It is a structural and functional prescription of an open, interoperable, horizontal system architecture for distributing computing, storage, control, and networking functions closer to the users along a cloud-to-thing continuum of communicating, computing, sensing, and actuating entities [13]. OpenFog reference architecture is particularly suited to IoT systems and facilitates deployments which highlight interoperability, performance, security, scalability, programmability, reliability, availability, serviceability, and agility [12].

OpenFog reference architecture is built on a set of core principles called pillars. These pillars form the belief, approach, and intent that guide the definition of OpenFog reference architecture and represent the key attributes of the systems [3].

1. Security: Security has many different descriptions and attributes such as privacy, anonymity, integrity, trust, attestation, verification, and measurement. These are key attributes for the OpenFog reference architecture. OpenFog architecture will enable the flexible creation of environments that address a broad spectrum of security concerns spanning from IoT devices to cloud and the fog layers in between.
2. Scalability: Scalability involves scalable performance, scalable capacity, scalable reliability, scalable security, scalable hardware, scalable software, and so on. It is essential for OpenFog architecture to adapt to workload, performance, system cost, business needs, and so on. Due to the variety of the application scenarios for fog computing, OpenFog architecture should enable elastic scaling of modest deployments through large mission critical deployments based on demand.
3. Openness: Openness includes composability, interoperability, open communication, location transparency, and so on. It is essential for the success of a ubiquitous fog computing ecosystem for IoT applications. OpenFog architecture

can support portability and fluidity of applications and services at instantiation, and enable the flexible discovery, collection, and redistribution of computing, network, and storage resources anywhere in the network.

4. Autonomy: Autonomy supports the autonomy of discovery, autonomy of orchestration and management, autonomy of security, autonomy of operation, cost saving, and so on. It is essential for the continue provision of functionality, and it is supported throughout the hierarchy. OpenFog architecture enables decision making to be made at all levels of a deployment's hierarchy including near the device or higher order layers.
5. Programmability: Programmability benefits adaptive infrastructure, resource efficient deployments, economical operations, enhanced security, and so on. It is the basis of openness, autonomy, and agility. OpenFog architecture supports highly adaptive programming at the software and hardware layers.
6. Reliability, Availability, and Serviceability (RAS): RAS has three main areas: hardware, software, and operations. It is resident throughout successful system architectures and takes on great importance in OpenFog architecture. OpenFog architecture will continue to deliver designed functionality under normal and adverse operating conditions and ensure continuous management and orchestration, and correct operation.
7. Agility: Agility focuses on transforming huge volume of data into actionable insights. It also deals with the highly dynamic nature of fog deployments and the need to respond quickly to change.
8. Hierarchy: Hierarchy involves many dimensions, including but not limited to devices in the hierarchy, monitoring and control in the hierarchy, operational support in the hierarchy, surrogacy in the hierarchy, and business support in the hierarchy. It is not required for all OpenFog architectures, but it is still expressed in most deployments. The resources of OpenFog reference can be seen as a logical hierarchy based on the functional requirements of an end-to-end IoT system.

As aforementioned, although hierarchy is not required for all OpenFog architectures, it is still expressed in most deployments. In most fog deployments, there are usually N-tiers of nodes. Figure 2.1 shows a subset of the combination of fog and cloud deployed to address various domain scenarios as framed by the layered view of IoT systems. Each fog element may represent a hierarchy of fog clusters fulfilling the same functional responsibilities. Depending on the scenario, multiple fog and cloud elements may collapse into a single physical deployment (the rightmost case). Each fog element may also represent a mesh of peer fog nodes in use cases like connected cars, electrical vehicle charging, and closed loop traffic systems. Generally speaking, nodes at the edge are typically focused on sensor data acquisition/collection, data normalization, and command/control of sensors and actuators. Based on the collected data, nodes in the next higher tier are focused on data filtering, compression, and transformation. They may also provide some edge analytics required for critical real-time or near real-time processing. Nodes at the higher tiers or nearest the back-end cloud are typically focused on aggregating data

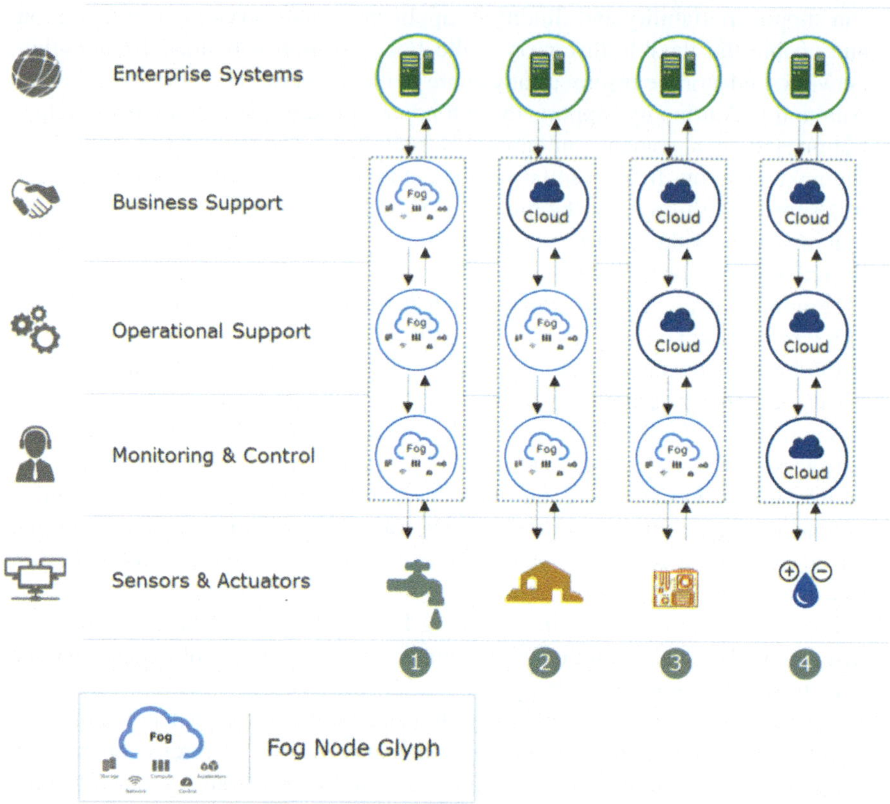

Fig. 2.1 IoT system deployment models [3]

and turning the data into knowledge [3]. This means that the multi-tier architecture uses a multitude of computational clients or edge devices, and it is important to note that the farther from the true edge, the greater the insights that can be realized.

OpenFog architectures offer several unique advantages, which are termed CEAL [2, 3].

1. Cognition: Awareness of client-centric objectives to enact autonomy. OpenFog architecture can be aware of customers' requirements and determine where to carry out the computing, network, and storage functions along the cloud-to-thing continuum.
2. Efficiency: Dynamic pooling of unused resources from nodes along the cloud-to-thing continuum. OpenFog architecture can discover, collect, and redistribute computing, network, and storage resources anywhere along this continuum to take full advantages of them.
3. Agility: Rapid innovation and affordable scaling under a common infrastructure. OpenFog architecture will make it easier to create an open market place for

individuals and small teams to use open software platforms and the proliferation of mobile devices to innovate, develop, deploy, and operate new services.

4. Latency: Real-time processing and cyber-physical system control. OpenFog architecture enables data analytics at the network edge and can support time-sensitive applications for local cyber-physical systems.

It is worth noting that OpenFog reference architecture should be thought of as complementary to, and an extension of the traditional cloud based architecture, where different levels of the cloud's functions can reside in multiple layers of the network's topology. Choosing between the cloud and OpenFog is not a binary decision. They are interdependent and mutually beneficial: certain functions are naturally more advantageous to carry out in fog while others are better suited to cloud. The segmentation of what tasks go to fog and what goes to the cloud are application-specific and could change dynamically based on the instantaneous state of the network. In conclusion, the traditional cloud will continue to constitute an important part of OpenFog architecture.

2.2.2 *Multi-Tier Computing Network*

Besides OpenFog reference architecture, a multi-tier computing network architecture integrating cloud computing, fog computing, edge computing, and sea computing has been proposed as a promising way to the development of IoT services [14]. Thus, a hybrid computing architecture is needed, which spans from the edge (fog) to the core (cloud). Multi-tier computing involves collaborations between cloud computing, fog computing, edge computing, and sea computing technologies, which have been developed for regional, local, and device levels, respectively [15].

IoT applications are increasingly intelligent due to more meaningful data, powerful processors, and sophisticated algorithms. Typical IoT applications are also shifting from simple data sensing, collection, and representation tasks towards complex information extraction and analysis. However, the applications usually follow rules and principles set by a specific industrial domain. Multi-tier computing resources, when integrated with environment cognition, big data, and AI technologies, could be used to develop a user-centric approach in which different IoT services are autonomously customized according to specific applications and user preferences.

Intelligent IoT networks and services with integrated, multi-tier computing resources can be thought of as a large company with a top-down, multi-tier organizational structure: managers and employees at different levels in the company have different resources, capabilities, and responsibilities in terms of data access and processing, task assignment, customer development, and decision making. Cloud computing is equivalent to the top hierarchical level in the company, possessing the most information sources, strongest analytical intelligence, maximum storage

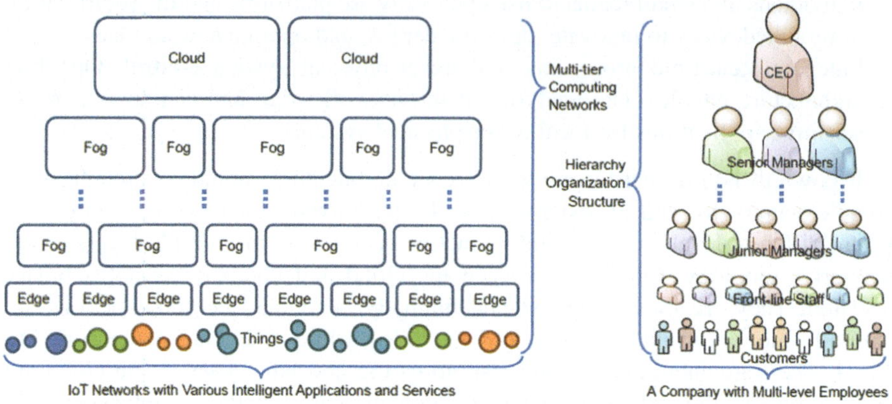

Fig. 2.2 Multi-tier computing network architecture, which integrates cloud, fog, edge, and sea computing technologies to enable intelligent services for anything at anytime and anywhere

space, and the highest decision making authority. As such, cloud computing is expected to handle challenging tasks at the global level, such as cross-domain data analysis and processing, abnormal behavior diagnosis and tracing, hidden problem prediction and searching, new knowledge discovery and creation, and long-term strategic planning and decisions, shown as Fig. 2.2. Edge computing, on the other hand, is equivalent to front-line staff, which have the least resources and capabilities but can directly interact with customers in different application domains. Therefore, edge computing is good at handling delay-sensitive tasks at the local level, such as data collection, data compression, information extraction, and event monitoring.

Between the cloud and the edge within the IoT network, there is fog computing, which is equivalent to mid-level management in the company. Like an efficient management system with many levels of resources, duties, and responsibilities, fog computing is a hierarchy of shared computing, communication, storage resources, and AI algorithms that can collaboratively handle complex and challenging tasks at the regional level, such as cross-domain data analysis, multi-source information processing, and on-site decision making for a large service coverage. Because user requirements are usually dynamic in terms of time and space, fog computing can provide a flexible approach to incorporate distributed resources at different geographical or logical locations in the IoT network, thus offering timely and effective services to any customers [16–18].

The devices, or things, of the IoT network are equivalent to the customers of the company, which have numerous different requests and demands for intelligent applications and services. Each device has limited processing, communication, storage, and power resources. But, collectively, they contribute to the concept of sea computing at the device level, which supports data sensing, environment cognition, mobility control, and other basic functions for individual things in real time [15].

To make the company a success, it is essential to have effective communication and collaboration mechanisms across different levels and units. Similarly, interaction and collaboration between cloud, fog, edge, and sea computing are vital in order to create an intelligent collaborative service architecture. As a result, this architecture actively connects shared computing, communication, and storage resources of all the nodes in the IoT networks, and fully utilizes their capabilities at different locations and levels, in order to provide intelligent, timely, and efficient services according to dynamic user requirements. Since most applications and their data do not require superior computing power, this architecture can significantly improve service quality and user experience while saving time, resources, and costs [14].

To make the above concepts reality, it will be necessary to provide a control and orchestration architecture to allocate computing, communication, storage resources. As a result, intelligent on-demand services can be created through the interoperability and integration of multi-tier computing sources in the IoT networks. To best meet the requirements of any user, centralized cloud computing with huge resources, secure environment, and powerful algorithms are needed, but also distributed fog and edge computing with shared resources, accessible environments, and simple algorithms for real-time decision making.

With heterogeneous computing resources and the collaborative service architecture, future multi-tier computing networks can offer densely distributed points for computation and storage, then effectively support a full-range of services in different environments and applications. For example, smart cars and drones require low-latency communications for receiving monitoring data and control messages. Autonomous driving and three-dimensional virtual reality games need powerful computing capabilities at device and edge levels, as well as broad communication bandwidths. Industrial IoT and smart city management systems demand extreme network reliability, data security, and service availability.

Fog computing is the bridge that connects centralized clouds and distributed edges of the network and thus plays a crucial role in managing multi-tier resources. It is, for example, responsible for vertical and horizontal interoperations, coordination of cross-domain collaborations, deployment of hierarchical rules and policies, and integration of multi-level services. Together with the edge, fog computing makes computing resources and intelligent services in the IoT networks more accessible, flexible, efficient, and cost-effective.

It should be noted that the development and standardization of multi-tier computing technologies have just begun and numerous technical challenges remain [3, 13]. Specifically, in terms of security, trust, and privacy, end users should be able to use a nearby shared computing node without compromising the protection of identity and personal data. Virtualization and resource visibility capabilities need to be developed to allow sharing of physical computing resources in different machines and hardware. Efficient orchestration, which refers to managing the allocation of heterogeneous resources and services, also requires interoperability between the nodes within the network. Finally, the design of an architecture that is able to serve various groups of users and is multi-tenant needs the development of a new business model that encourages an ecosystem of shared resources and collaborative services. These issues necessitate comprehensive research efforts using multi-disciplinary approaches across different technological and economic fields [11, 19, 20]. The detailed description of the challenges will be discussed in the following subsection.

2.3 Fog Computing Technologies

2.3.1 Networking Technologies

In most fog deployments, networks facilitate collaboration among layers and communication within the fog nodes to sensors and up to higher levels in the hierarchy up to and including the cloud. Therefore, thing-to-edge, edge-to-fog, fog-to-fog, and fog-to-cloud interconnection and communication are essential for the fog computing architectures and deployments. According to the node's purpose and location, they can be connected with each other using a wired network or a wireless network.

For wired connectivity, there are many standards, types, and interfaces for physical connectivity that can be utilized to connect a fog node. Ethernet links supported on copper or fiber links typically support speeds ranging from 10 Mbps to 100 Gbps. For connections requiring higher speeds and longer reach, optical fiber cables may be used. For connecting a fog node to IoT devices or sensors, there are a variety of non-Ethernet protocols can be used. For example, in an industrial environment, the fog node may be required to support a Controller Area Network (CAN) bus or other fieldbus standards for communicating with lower layer applications and processes. For industrial automation uses, guaranteed data delivery is critical. This type of networking (usually using Ethernet) is called Time Sensitive Networking (TSN) also known as Deterministic Ethernet. TSN uses standards-based time synchronization technology (e.g., IEEE 1588) and bandwidth reservation (class-based QoS) to prioritize control traffic in a standard Ethernet environment [3].

For wireless connectivity, it can be classified into three major areas: wireless WAN (WWAN), Wireless LAN (WLAN), and wireless personal area networks (WPAN). WWAN technologies are used when large geographic area coverage is required. A variety of protocols and standards exist, including cellular technologies (such as 3G, 4G LTE, and 5G), narrow band IoT (NB-IoT), low-power wide area networks (LPWAN), and so on. WLANs utilize a variety of topologies and protocols, but WLAN has become synonymous with Wi-Fi. WLANs are a good choice for smaller geographical areas, often within a building or campus. WPAN is characterized by a short communication range, low power consumption, and low cost. WPANs may be used with wearable devices and home management systems. The most common WPAN technologies are Bluetooth, infrared (IR), ZigBee, Z-Wave, and IEEE 802.15.4 (Low Rate WPAN). Besides, near field communication (NFC) may be used when fog nodes support devices that need to communicate in very close proximity [3].

2.3.2 Computing Technologies

As more and more data is generated and processed near end users under stringent delay requirements, this will increasingly increase the computational requirements of equipment, especially the edge devices. Additionally, as AI techniques are to be widely deployed to realize the autonomy and agility of fog computing, having general purpose computation at the edge will continue to be important. Typically, compute is likely to be implemented as one or more multi-core Central Processing Unit (CPU). In addition to traditional CPUs, some fog nodes, especially those engaged in enhanced analytics, require CPU throughput in excess of what can be economically (power or processing efficient) provided by standard current server and enterprise CPU chips. In these cases, accelerator modules will be configured next to the processor modules (or tightly integrated) to provide supplementary computational throughput. Among them, Graphics Processing Unit (GPUs) and Field Programmable Gate Arrays (FPGAs) are the most common.

GPUs often contain thousands of simple cores. For applications that can efficiently exploit their massive parallelism, they can be an order of magnitude faster and provide significant power and space savings. Multiple GPUs can be equipped on each standard CPU. However, to achieve this capability power delivery and physical and electrical connection to the node would need to be increased which can increase the overall node power consumption. FPGAs are large collections of gate-level programmable hardware resources. They can be configured with custom logic designs to solve very specific problems very efficiently. However, depending upon deployment the additional power reduction compared with other accelerators may require additional lower level knowledge (e.g., VHDL). In many cases FPGA provide better power efficiency compared with discrete GPUs [3].

2.3.3 Storage Technologies

As fog nodes collect and process data across the hierarchy, storage tiers which are typically only seen in data centers will emerge on fog nodes. Thus, different types of storage will be required in fog nodes. The most common ones include RAM arrays, solid state drives, and fixed spinning disks.

As data is created from sensors, the node will need to operate on that data in close to real-time operation. RAM arrays satisfy this requirement versus additional latency when accessing non-volatile storage. Many fog nodes will also have on-package memory to satisfy the latency aspects required for certain scenarios. Solid state drives, or flash-based storage may be used for the majority of fog applications because of its reliability, IOPs, low power requirements, and environmental robustness. These include PCIe and SATA attached SSDs. Additionally, new classes of solid state media are beginning to emerge with new programming models. These include 3DXpoint and NVDIMM-P. For large, cost sensitive storage applications, fog nodes may contain rotating disks (fixed spinning disks), sometimes arranged as Redundant Array of Inexpensive Disks (RAID) arrays [3].

For fog computing, storage devices should support encryption and key management and authentication by supporting standards such as AES-256 and TCG Opal etc. Besides, in virtualized fog computing architectures, the storage devices should also support ID-based performance allocation by providing adjustable storage resources (IOPS or bandwidth) to specific applications. Finally, supporting data encryption, real-time information, early warnings about the health of the media, and self-healing properties are also important in most fog deployments.

2.3.4 Virtualization Technologies

According to the fog computing architectures described above, fog computing architectures are highly virtualized ones. Under such architectures, processing, accelerators, storage and networking functions should all be virtualized to maximize the efficiency, flexibility, and manageability of the fog system. Virtualization may also incorporate aspects of containerization, depending upon the software layers that run on the hardware. Therefore, both hardware virtualization technologies and software virtualization technologies are needed through the infrastructure.

Hardware-based virtualization mechanisms are available in almost all processor hardware that would be used to implement fog platforms. It may also play an important role in system security. Hardware virtualization for I/O and compute enables multiple entities to share the same physical system. Virtualization is also very useful in ensuring that VMs may not utilize instructions or system components that they are not by design supposed to utilize. Besides hardware-based virtualization mechanisms, containerization technologies are also used in fog computing to help with isolation. Containers may offer a lower weight isolation mechanism within a fog computing environment. The isolation guarantees are only made by the OS and not fully based in silicon. This shifts the isolation requirements from the silicon to the software running on the silicon. The decision to use containers or VMs for isolation are usually based on security considerations for a given use case [3].

To provide hardware-based virtualization support and OS based isolation support for running software and application microservices, software virtualization technologies are necessary. With the increasing use of Software Defined Networking (SDN) implementations to replace dedicated devices, these "appliances" are increasingly being implemented as software solutions in virtual machines and Linux containers. Software containers provide a good mechanism for fine-grained separation of applications and microservices running on the software backplane. A software container, unlike a VM, often does not require or contain a separate OS. A container uses resource isolation of the CPU, memory, block I/O, network, etc., and separate namespaces to isolate the application's view of the operating system. Within a container, applications can be configured, resources isolated, and services restricted. Multiple containers share the same kernel, but each container can be constrained to only use a defined amount of resources, such as CPU, memory, or I/O. Containers facilitate highly distributed systems by allowing multiple applications to run on a single physical compute node, across multiple VMs, and across multiple physical compute nodes [3].

2.3.5 Artificial Intelligence

In most fog computing deployments, AI techniques are utilized to handle the sophisticated analytics tasks and optimize the system operations. Taking smart buildings as an example, each floor, wing, or even individual room could contain its own fog node and form a hierarchical control system. Each fog node could perform emergency monitoring, security functions, climate and lighting control and provide a compute and storage infrastructure for building residents to support smartphones, tablets, and desktop computers. Locally stored operational history can be aggregated and sent to the cloud for large-scale analytics. These analytics can be applied to artificial intelligence to create optimized models, which are then downloaded to the local fog infrastructure for execution [3].

AI is in the forefront of research today. Generally, AI techniques or machine learning methods can be grouped into three categories: supervised learning, unsupervised learning, and reinforcement learning. Supervised learning is a type of system in which both input and desired output data are provided. Input and output data are labeled to provide a learning basis for future data processing. Supervised learning mainly includes classification and regression. In contrast, unsupervised learning is a method used to enable machines to classify both tangible and intangible objects without providing the machines any prior information about the objects. The most common unsupervised learning is clustering. Reinforcement learning allows the machine or software agent to learn its behavior based on feedback from the environment, in order to maximize its rewards.

Conventional machine learning techniques are limited in their ability to process natural data in their raw form. Constructing a machine learning system usually requires considerable engineering expertise to design a feature extractor. With the efforts of Geoffrey Hinton, Yoshua Bengio, and Yann LeCun, deep learning has been promoted to solve the problems of conventional machine learning techniques and make a great success. Deep learning is a method that allows computational models composed of multiple processing layers (termed neural networks) to learn representations of data with multiple levels of abstraction. These methods have dramatically improved the state-of-the-art in speech recognition, visual object recognition, object detection, and many other domains. Convolutional neural networks (CNN) are designed to process data that come in the form of multiple arrays, for example, a color image composed of three 2D arrays containing pixel intensities in the three color channels. While for tasks that involve sequential inputs, such as speech and language, it is often better to use recurrent neural networks (RNNs). A variation of

RNN is the long short-term memory (LSTM) networks, which use special hidden units to remember inputs for a long time. LSTM networks are specially suited to deal with long-term time series data.

2.4 Applications and Challenges

2.4.1 Collaborative Robot System

Robots are popular now and have been entered people's lives and industries in many fields. They bring us much convenience in daily life, save huge manpower in factories, and complete tasks mission-impossible for human beings. In the real world, robots are often needed to explore a prior-unknown environment. For example, when people are buried in a collapsed building in an earthquake, and the space/condition does not allow rescue person/animal to enter, an advanced method is to send a suitable size and shape robot to detect the location of the lives, environment, and the living conditions like oxygen level and temperature, etc., mapping information is critical to rescue planning. This requires the explorer robot to construct a map of the space. Meanwhile, to do so it needs to know its location and orientation, too. In robotics, to perform concurrent construction of a map of the environment and estimation of the robot state is simultaneous localization and mapping (SLAM).

Generally, in above rescue use case, robot SLAM needs to be low cost, low-power consumption, accurate, and speedy. However, these requirements restrict each other mutually. First, tens or hundreds of rescue robots may be used in one saving campaign so robot cost is a big concern. Popular SLAM sensors include laser radars and cameras; however, although generally laser radars perform better in accuracy, they are much expensive than usual cameras, so the latter is more suitable for massive rescue robots. Second, accurate mapping and localization are very important to rescue path planning, to achieve this it needs high-performance computing unit, especially at the step of optimization that involves many advanced algorithms, and this of course, is contradictory to the low-cost requirement aforementioned. Third, rescue robots use battery; hence, battery life is a key consideration. To reduce computing tasks of robots, e.g., to use low-performance algorithms is an effective way to save robot energy consumption; however, it may lead to inaccurate SLAM. Fourth, as all know, time is critical in many rescue cases; hence, it requires robots to move as quick as possible and perform fast SLAM, imposing much pressure to the onboard computing unit. Similarly, low-complexity algorithms may be used to save SLAM time; however, a sequence is that the SLAM accuracy may be affected. Fifth, in a large area where multiple robots are used, it requires collaborations between the robots in SLAM and to merge the maps finally, thus a network is needed and one robot needs to be a leader in SLAM and to merge maps, this leader robot, of course, will consume additional energy. However, network is usually not available in disaster scenarios.

An effective solution to robot SLAM issues is fog computing. OpenFog has identified eight pillars that a fog network needs to consider. They are also challenges and requirements faced by fog-enabled robot SLAM.

- **Security**: Robot SLAM enabled by fog has privacy critical, mission critical, and even life critical aspects, particularly in rescue applications. Connecting robot to a fog network and cloud exposes both robots and SLAM process to possible hacking and may suffer loss of map data and/or tampered SLAM process.
- **Scalability**: It needs to address dynamic technical and business needs behind deployment of fog network for robot SLAM, meaning the performance, capacity, security, reliability, hardware, and software of the fog network should be scalable. A FN may scale its internal capacity though addition of hardware and software, and a fog network may scale up and out by add new node if computing tasks are too heavy.
- **Open**: Openness is essential to the success of a ubiquitous fog network. It needs diverse vendors for consideration of cost, quality, and innovation. With interoperability and software technologies, fog computing can be deployed or defined at anywhere. A robot can perform fog-enabled SLAM when it has connection to a network nearby with required software.
- **Autonomy**: It requires the fog network can assist SLAM when its external service or connection is loss. This eliminates the need of a centralized processing, e.g., at cloud. In disaster scenarios where cloud connection is usually not available, it requires the fog network could perform SLAM as well. If any working FN is lost, then the remaining FNs need to take over the tasks and continue SLAM processes.
- **RAS**: Reliability is essential to fog-enabled robot SLAM. The network, software, hardware need to be all reliable. Rescue robots are usually deployed in harsh environments like earthquake, fire, storm, etc., very robust communication is required. Availability, measured by uptime, is required for continuous support to robot SLAM and serviceability, meaning correct operation of the fog-enabled robot SLAM, imposes requirement of self-healing, autonomous configuration, etc.
- **Agility**: It needs the fog network response quickly when robots moving in the environment, or the environment is dynamic and changes fast. In the case of rescue robots saving lives, it is critical for the fog networks to map and position as fast as it can.
- **Hierarchy**: When multiple robots are used to building map jointly, then a master FN merges the maps from different robots. A master FN is a higher layer node comparing to other FNs and may coordinate other FNs behavior. Moreover, if there are other applications associated with SLAM, like rescue, then SLAM function may need to interact with higher layer services like path planning, monitoring, business operation, etc.
- **Programmability**: SLAM may dynamically change with environment, network topology, and so on, so the fog network and nodes may highly adaptively program at hardware and software layer. The computing FN or the group FNs may be re-tasked automatically to accommodate the changes.

2.4.2 Wireless Communication Network

As one fundamental cornerstone of the global digital economy and our connected society, wireless communication networks are and will continue to provide diverse services for people and things. Meanwhile, we see the wireless networks are also experiencing an unprecedented traffic growth and are expected to offer an increasing variety of services and applications with different traffic patterns and QoS and quality of experience (QoE) requirements. For example, some services demand ultra-high data rates, while some necessitate ultra-high reliability and ultra-low latency. The popular smart user devices, such as the smart phones, the laptops, and the tablets, are now pushing the current wireless networks to their limits. To cope with the continuing traffic growth and service expanding, future wireless networks will have to be heterogeneous and densely deployed, featuring the coexistence of different radio access technologies (RATs) including LTE/LTE-Advanced, Wi-Fi, IoT, 5G new radio (NR), etc.

Accordingly, future wireless networks will be significantly more complex to deploy and operate than the existing 3G/4G mobile networks, due to the dense deployment of small BSs and the heterogeneities of network nodes, RATs, and services (and hence traffic patterns). The increasing management complexity of wireless networks has made it evident for the necessity of wireless network self-optimization, where wireless networks are automated to minimize human intervention and to proactively optimize network deployment, operation, and multi-RAT resource allocation to meet increasing service demand from people and things.

The concept of self-organizing networks (SONs) was first introduced in 3GPP Release 8 with the aim to automate the operation and management of wireless networks. SON functionalities can be generally classified into self-configuration, self-healing, and self-optimization. In subsequent 3GPP releases, SON concepts and technologies have been extensively developed. Nowadays, SON technologies are widely expected to:

1. enhance the intelligence and autonomous adaptability of wireless networks, especially the RANs;
2. reduce the capital and operational expenses (CAPEX and OPEX) for wireless network operators;
3. improve both network-wide performance and user specific QoS and QoE.

Moreover, future wireless networks will require (near) real-time and scalable SON solutions that are tailored for multi-vendor and multiple RAT networks, as well as for M2M communications and IoT systems.

Wireless network self-optimization comprises various mechanisms that optimize network parameters during operation according to measurement data taken at different parts of the network. The number of network parameters that need to be self-optimized are still increasing and would be enormous in the near future.

Nevertheless, most existing SON solutions are mainly based on heuristics, with the automated information processing limited to relatively simple methods. Many open challenges of SON remain unsolved. For example, the automated coordination between different SON functions and the trade-off between centralized and distributed SON implementations are still open research problems.

Fog computing has been considered as a promising paradigm shift to enable autonomous management and operation of wireless networks. Fog computing features a distributed computing infrastructure, its proximity to the network edge and end users, and the dense geographical distribution of FNs. These allow fog computing to exploit the local signal processing and computing, cooperative resource management, and distributed storing/caching capabilities at the network edge.

A FN is typically a virtualized platform hosted on either a dedicated computing node, which is equipped with communication interface(s), or a networking node/device, such as a BS, an AP, a router, or a switch. This asks for the combination of fog computing and two other emerging technologies: SDN and network function virtualization (NFV). SDN implies a logically centralized network control plane, which allows the implementation of sophisticated mechanisms for traffic control and resource management. In SDN, the control plane carries signaling traffic, calculates routes for data flows, and performs configuration and management for the network, while the data plane is responsible for transporting data packets. NFV implements network functions in software that can run on a range of industry standard server hardware and that can be moved to or instantiated in various locations of the network as required, without the need to install new equipment. Fog computing, in conjunction with SDN and NFV, can bring extensive programmability and flexibility into wireless networks, and thus enable distributed and intelligent SON functionalities in (near) real-time.

In a fog-enabled wireless network, a large amount of signal processing and computing is performed in a distributed manner, and local data can be stored and processed in edge devices, such as BSs, APs, and user devices, thus providing support for applications that require very low and/or predictable latency and offering mobility support, geo-distribution, location awareness, and low latency [21]. For instance, mobile application processing delays can be reduced by offloading the associated computationally intensive tasks to the FNs that are close to the corresponding mobile applications.

We envision that the fog computing paradigm will allow the self-optimized computing, control, caching, storage, and networking functions to be dynamically relocated among the cloud, the fog, the network edge, and the things, as well as allow self-optimized management of network function and service lifecycles. Accordingly, fog computing will lead to new opportunities in the design of SON for wireless networks, by exploring the various trade-offs between distributed and centralized operation, between local and global optimization, etc.

2.4.3 Intelligent Transportation System

Currently, transportation systems are an indispensable part of human activities. As people become more dependent on transportation systems, the transportation systems themselves are facing not only many opportunities, but also lots of challenges, such as traffic congestion, accident risks, and efficient transportation management. Intelligent transportation system (ITS) is proposed to establish a real-time, accurate, and efficient transportation management system, using information, communication, control, computer technology, and other current technologies. ITS is an integrated system of people, roads, and vehicles, utilizing a variety of advanced technologies in communication, automation, computing, etc. An ITS uses the data collected from various sources, such as cameras, sensors, global positioning system (GPS) receivers, and other vehicles, to optimize the system's performance in terms of traffic flow, safety, delay, and fuel consumption. As the number of vehicles grows rapidly and vehicles become more autonomous, ITS will face more critical challenges, which can be summarized as follows:

- **Explosive computing workload**
  With the popularity of autonomous driving (AD) technology, the vehicles are expected to create significant amounts of data by 2020. An AD vehicle will generate and consume roughly 40 terabytes of data for every 8 h of driving [2]. The huge amount of data brings tremendous burden to the in-vehicle computing processors due to the limitation on power, space, and heat dissipation. One solution is to upload the computing workload to cloud, which has powerful computing and storage capabilities. But the data transmission between vehicle and cloud comes at the expense of network resources, while the real-time requirements cannot be satisfied because of the limited fronthaul capacity. Therefore, it is necessary to introduce a new architecture to deal with the explosive computing workload with low latency and cost.
- **Ultra-reliable and low-latency communications**
  To guarantee the transportation safety, especially for AD vehicle, the vehicular communication systems have extremely high requirements on reliability and transmission latency. For ultra-reliable and low-latency communications (URLLC), a general reliability requirement for one transmission of a packet is 1-10-5 for 32 bytes with a user plane latency of 1 ms. This requirement is a great challenge to the existing RAT. On the other hand, when the vehicle is running in regions where there is no AP with cellular connection, the connection between vehicle and network may be broken off and the vehicle has to rely on expensive data communication over satellite. Thus, novel RAT and network architecture need to be designed to satisfy the reliability and latency requirements, as well as providing continuous connectivity to vehicles.
- **Security and privacy**
  ITS has strict security and privacy requirements on computing, communications, and many other parts of platform architecture. Security and privacy considerations in ITS include:

  - Safe Driving: This is highly coupled with vehicles sensing capabilities and the ability to adapt the automated driving based on variable road conditions and situations sensed locally.
  - Platform Security: This necessitates functional safety features in the in-vehicle platforms, the road side units (RSUs), and the network platforms.
  - Secure Communication: This requires mutual authentication between all pairs of communicating parties as well as powerful encryption mechanisms.
  - Trust and Trustworthiness: This includes trust between pairs of communicating nodes as well as passengers' trust of AD vehicles, which creates the need for new in-vehicle experiences.
  - Privacy: Assurance that the data generated about the vehicle, its owners, its passengers, and location can meet privacy preservation requirements, which may be established early in the data lifecycle, for example, by policies attached to the data, the system, and/or conditioned on the context.

- **Data sharing across operational silos**
  In order to alleviate traffic congestion and avoid traffic accidents, transportation information, such as road conditions, driving status, and traffic information, should be shared in time across different vehicles and infrastructures. However, due to the differences of protocols, interfaces, and implementations, the devices from different operators are difficult to share data with each other, which forms the so-called operational silos. To improve the performance of ITS, a general platform should be provided to the devices from different operators, which allows data sharing across the operational silos.

Fog computing provides a critical architecture for today's connected world as it enables slow latency, high reliable, and high efficient operations, as well as provides strong support for mobile applications. Based on the above features, fog computing enables the critical functions of ITS by collaborating, cooperating, and utilizing the resources of underlying infrastructures within roads, smart highways, and smart cities. Fog computing will address the technical challenges in ITS and will help scale the deployment environment for billions of personal and commercial smart vehicles.

- **Security**: Fog requires every FN in the cloud-to-thing continuum to have high-assurance security mechanisms, which can provide on-demand security services to resource-constrained devices as well as offer trustworthy information processing, storage, and transport throughout the fog-enabled ITS.
- **Scalability**: The scalability of fog architecture includes the hardware or software adaptation of individual FNs, the addition of FNs in the fog network, and the storage, network, and other services scaled with the fog infrastructure. Besides, the breakdown of operations between in-vehicles, RSUs, infrastructures, and cloud eases scalability, management, and orchestration.
- **Openness**: For fog-enabled ITS, interoperability and data sharing take place at different phases and layers. Vehicle manufacturers, fleet owners, insurance

companies, and government regulators all participate to establish an open multi-party ecosystem to improve their cost, functionality, safety, and rate of innovation.

- **Autonomy**: Each FN from vehicle to cloud has certain capabilities of processing, storage, and decision making. Vehicles and RSUs are autonomous and can perform critical functions without the assist of cloud resources.
- **RAS**: Having each vehicle, RSU with compute and storage, and access network infrastructure with compute and storage as a FN can ensure RAS in the ITS. In addition, a group of vehicles can act as federated resources for compute and path planning, which ensures RAS even if there is no connectivity to the access network (e.g., in rural areas or in a tunnel).
- **Agility**: Each FN from vehicle to cloud can agilely make immediate decision. Pushing the dynamic location update schemes from the cloud to the lower layer FN will greatly increase the agility of the system by enabling real-time reactions to location-based traffic patterns or services provided. Agility allows the fog network to adapt quickly to changing conditions or changing customer needs.
- **Hierarchy**: Smart vehicles require connectivity along the fog hierarchy to distribute computing workload and data storage. This takes place through thing-to-fog, fog-to-fog, and fog-to-cloud. This hierarchy will greatly improve the paging resolution from the cloud, as a vehicle may be handed over between RSUs at a rapid pace, and those RSUs will connect to a single infrastructure that maintains a connection to the cloud.
- **Programmability**: Programmability is at the heart of a fog-enabled ITS and pervades all operations involving computing, storage, communications, and layering through a fog hierarchy. Federation of computing, storage, and communication enable dynamic adjustment of platform capabilities depending on needs.

2.4.4 Smart Home

Smart home is a house system that incorporates advanced automation systems to provide the inhabitants with sophisticated monitoring and control over the building's functions. The UK Department of Trade and Industry (DTI) came up with the following definition for a smart home: "A dwelling incorporating a communications network that connects the key electrical appliances and services, and allows them to be remotely controlled, monitored or accessed." For example, a smart home may control lighting, temperature, multi-media, security, window, and door operations, as well as many other functions. Smart home is a typical intelligent IoT system, where each smart appliance/device is able to connect to the Internet and carry out some computing tasks. Each appliance/device can be viewed as an IoT node and they form a local network.

A smart home may contain thousands of sensors to measure various building operating parameters, including temperature, humidity, door open/close, security, and air quality. These sensors capture telemetry data at various intervals and transmit this information to cloud or a local storage server. Once this information is processed or analyzed, controller-driven actuators will adjust building conditions as necessary. However, it is difficult and inefficient to transmit all the data to the cloud. For example, some visual data from camera sensors, which is huge in size, is impractical to be transmitted to the cloud to obtain real-time insights. Besides, some of the processing and response of the data is extremely time-sensitive. For example, turning on fire suppression systems in response to a fire event or locking down an area if an unauthorized person tries to gain entry. Therefore, to guarantee efficient management and real-time response, processing in close proximity to the infrastructure devices is required.

Fog computing provides a beam of light to the implementation of smart home. Using the hierarchical design of the fog computing, each floor, wing, or even individual room could contain its own fog node and create a hierarchical control system. Each fog node could be responsible for performing emergency monitoring and response functions and building security functions, and controlling climate and lighting. They could also provide a more robust compute and storage infrastructure for building residents to support sensors, smartphones, and desktop computers. Local sensors can first transmit sensing or monitoring data to local fog nodes. Fog nodes will preprocess these data and make simple analysis. For some time-sensitive tasks, fog nodes will make quick response to actuators. Otherwise, the preprocessed data can be aggregated and sent to the cloud for large-scale analytics. These analytics can be applied to machine learning to create optimized models, which are then downloaded to the local fog infrastructure for execution.

2.5 Conclusion

By selectively moving compute, storage, communication, control, and decision making along the cloud-to-thing continuum closer to the network edge, fog computing has been promoted to solve the key challenges of future intelligent IoT systems. In this chapter, we have discussed fog computing architectures and technologies. More specifically, we first introduced fog computing architectures in industry and academia, termed OpenFog reference architecture and multi-tier computing network, respectively. Then, we discussed the enabling technologies for fog deployments, including networking technologies, computing technologies, storage technologies, virtualization technologies, and AI. Finally, we described some detailed use cases of fog computing and look into how they would benefit from fog computing and what challenges would exist as well.

References

1. Verma S, Kawamoto Y, Fadlullah ZM, Nishiyama H, Kato N (2017) A survey on network methodologies for real-time analytics of massive IoT data and open research issues. IEEE Commun Surv Tutorials 19(3):1457–1477
2. Chiang M, Zhang T (2016) Fog and IoT: an overview of research opportunities. IEEE Internet Things J 3(6):854–864
3. The OpenFog Consortium (2017) OpenFog reference architecture for fog computing. https://www.openfogconsortium.org/ra/
4. Vaquero LM, Rodero-Merino L (2014) Finding your way in the fog: towards a comprehensive definition of fog computing. SIGCOMM Comput Commun Rev 44(5):27–32
5. Bonomi F, Milito R, Zhu J, Addepalli S (2012) Fog computing and its role in the internet of things. In: Proceedings of the first edition of the MCC workshop on mobile cloud computing. ACM, New York, p 13–16
6. Bader A, Ghazzai H, Kadri A, Alouini MS (2016) Front-end intelligence for large-scale application-oriented internet-of-things. IEEE Access 4:3257–3271
7. Satyanarayanan M, Bahl V, Caceres R, Davies N (2009) The case for VM-based cloudlets in mobile computing. IEEE Pervasive Comput 8(4):14–23
8. Barbarossa S, Sardellitti S, Lorenzo PD (2014) Communicating while computing: distributed mobile cloud computing over 5G heterogeneous networks. IEEE Signal Process Mag 31(6): 45–55
9. Mach P, Becvar Z (2017) Mobile edge computing: a survey on architecture and computation offloading. IEEE Commun Surv Tutorials 9(3):1628–1656
10. Hu YC, Patel M, Sabella D, Sprecher N, Young V (2015) Mobile edge computing: a key technology towards 5G. 11, 1st edn
11. Mouradian C, Naboulsi D, Yangui S, Glitho RH, Morrow MJ, Polakos PA (2018) A comprehensive survey on fog computing: state-of-the-art and research challenges. IEEE Commun Surv Tutorials 20(1):416–464
12. The OpenFog Consortium (2016) Openfog architecture overview. https://www.openfogconsortium.org/ra/
13. IEEE Standard 1934–2018 (2018) IEEE standard for adoption of OpenFog reference architecture for fog computing. https://standards.ieee.org/standard/1934-2018.html
14. Chen N, Yang Y, Zhang T, Zhou MT, Luo X, Zao JK (2018) Fog as a service technology. IEEE Commun Mag 56(11):95–101
15. Jiang MH (2010) Urbanization and cyberization: historical opportunity for China's development. Expo 2010 Shanghai China forum.
16. Yang Y, Wang KL, Zhang GW, Chen X, Luo X, Zhou MT (2018) MEETS: maximal energy efficient task scheduling in homogeneous fog networks. IEEE Internet Things J 5(5):4076–4087
17. Yang Y, Zhao S, Zhang W, Chen Y, Luo X, Wang J (2018) DEBTS: delay energy balanced task scheduling in homogeneous fog networks. IEEE Internet Things J 5(3):2094–2106
18. Zhao S, Yang Y, Shao Z, Yang X, Qian H, Wang CX (2018) FEMOS: fog-enabled multi-tier operations scheduling in dynamic wireless networks. IEEE Internet Things J 5(2):1169–1183
19. Byers CC (2017) Architectural imperatives for fog computing: use cases, requirements and architectural techniques for fog-enabled IoT networks. IEEE Commun Mag 55(8):14–20
20. Wen Z, Yang R, Garraghan P, Lin T, Xu J, Rovatsos M (2017) Fog orchestration for IoT services. IEEE Internet Comput 21(2):16–24
21. Natarajan P, Bonomi F, Milito R, Zhu J (2014) Fog computing: a platform for internet of things and analytics. In: Big data and internet of things: a roadmap for smart environments. Springer, Berlin, pp 169–186

Chapter 3
Analytical Framework for Multi-Task Multi-Helper Fog Networks

3.1 Introduction

Fog computing [1–3] has become an attractive technology to support delay-sensitive IoT applications [4–6]. Different from the centralized cloud computing, fog computing distributes the shared and flexible communication, computation, and storage resources along the continuum from cloud to things [2]. Specifically, the customized or pre-existing end, edge, and access equipment along the cloud-to-things continuum, which is collectively called fog node (FN), can form a resource pool and share resources with each other. As a result, more delay-sensitive applications can be served by various neighboring FNs with diverse resources and capabilities, instead of remote cloud servers. Benefiting from such a "Fog as a Service Technology" (FA^{2}ST) trend [7], the service efficiency can be efficiently improved, and the service delay will also be greatly reduced.

Although fog computing shows great potential advantages over legacy cloud computing, it also poses huge challenges in many aspects, especially in task offloading and resource allocation, which are key to reap the full benefits of fog computing [8]. To be specific, in a typical heterogeneous fog network, numbers of varied FNs are commonly distributed in different geographical locations. Due to the large scale, geographic dispersion, and node heterogeneity, the computational complexity may grow exponentially with the network size, and the global information may be difficult to obtain, for the conventional centralized methods. Thus, the self-organizing distributed methods, which require low complexity and local information, are preferred. Besides, the way that FNs make decisions by themselves can balance each FN's interest and stimulate them to participate. Taking the general heterogeneous fog network in Fig. 3.1 as an example, it consists of many randomly distributed FNs with diverse resources and capabilities. Some FNs have delay-sensitive tasks to process, and thus they are called task nodes (TNs). While some have spare resources to help neighboring TNs to process tasks, and thus they are called helper nodes (HNs). How to effectively map multiple tasks

Y. Yang et al., *Fog-Enabled Intelligent IoT Systems*,
https://doi.org/10.1007/978-3-030-23185-9_3

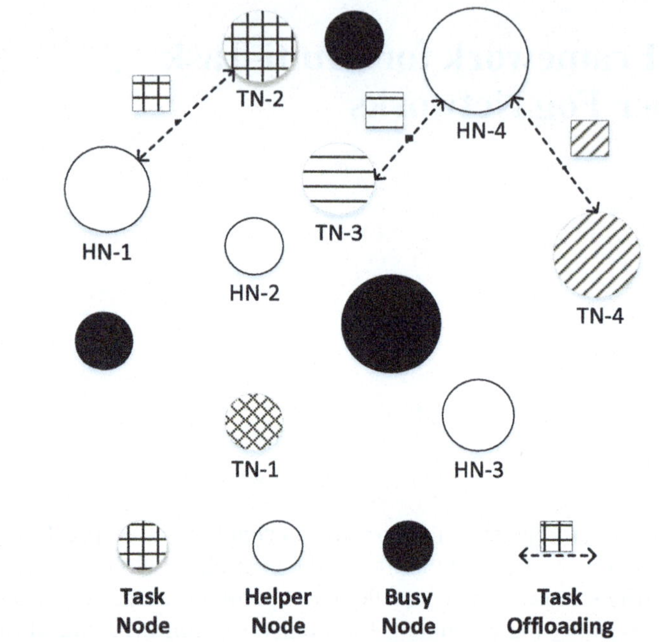

(a)

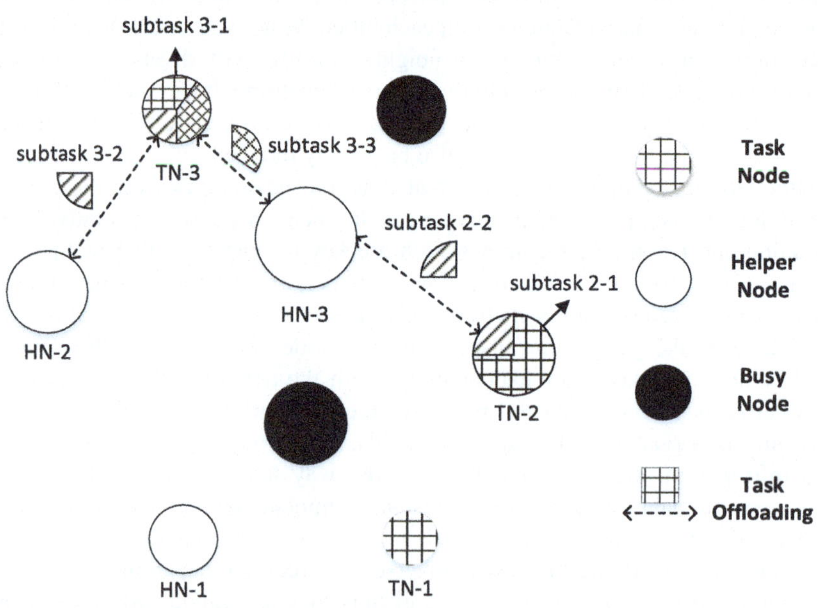

(b)

Fig. 3.1 A general heterogeneous MTMH fog network with TNs, HNs, and BNs. They are with different computation resources and capabilities, which are distinguished by the size of cycles. (**a**) Non-splittable tasks. (**b**) Splittable tasks

or TNs into multiple HNs to minimize every task's delay in a distributed manner is a fundamental challenge, which is called the Multi-Task Multi-Helper (MTMH) problem [9]. It is worth noting that such a MTMH network and the corresponding problem is quite general and can be applied in different application scenarios. For fog-enabled robot systems, TNs can be the robots, and HNs can be the idle robots, edge nodes in the room, factory. For fog-enabled wireless communication networks, TNs can be the mobile users, and HNs can be the AP, BS in the vicinity. For fog-enabled intelligent transportation systems, TNs can be the vehicles, and HNs can be the roadside units around. For fog-enabled smart home, TNs can be the heterogeneous sensors, devices, and HNs can be the routers, local edge nodes in the house.

Many previous works have investigated the task offloading problem in fog networks with a single TN and multiple HNs [10–12], or multiple TNs but a single HN [13–17]. Meng et al. [10] gave a closed-form solution to the energy-minimization workload allocation problem with delay constraints. Dinh et al. [11] proposed an optimization framework of minimizing both the execution delay and energy consumption by jointly optimizing the task allocation decision and device's CPU frequency. Yang et al. [12] developed an analytical model for maximizing the overall energy efficiency in homogeneous fog networks by considering the task allocation and time slots allocation. Nowak et al. [13] investigated the delay-minimization task offloading problem in a time division multiple access (TDMA) based fog network and calculated a closed-form solution to the workload allocation. Zhao et al. [14] considered an energy-minimization task offloading problem by jointly optimizing the offloading selection, radio resource allocation, and computation resource allocation and proposed two algorithms based on the Branch-and-Bound method and greedy method, respectively. Chen et al. [15] studied an overhead-minimization task offloading problem in a multi-channel wireless interference/contention environment and designed an efficient distributed algorithm to decide the offloading decision and channel allocation.

Notably, increasingly more attention is devoted to the MTMH task offloading problem [18–24]. Mansouri et al. [18] and Yang et al. [19] studied the overhead-minimization task offloading problem in hierarchical fog-cloud networks and edge computing empowered small-cell networks, respectively. Tran et al. [20] and Pham et al. [21] investigated how to jointly schedule tasks and allocate resources to maximize the utility and minimize the overhead in multi-cell edge computing networks, separately. Li et al. [22] considered how to minimize the system energy consumption in a novel non-orthogonal multiple access (NOMA) based edge computing network, by jointly scheduling tasks and allocating resources. Zhao et al. [23, 24] analyzed the delay-energy tradeoff in task offloading and resource allocation for homogeneous and heterogeneous fog networks, respectively.

The existing works [23–29] studied the MTMH problems in fog networks by considering different objectives from diverse perspectives. However, most of them [23–28] were devoted to optimizing the overall system performance, using centralized methods. Nonetheless, the centralized methods with system objectives may be faced with several challenges, especially in large-scale fog networks with heterogeneous and dispersive FNs.

- Due to the heterogeneity of FNs and with the enhancement of capabilities, FNs may make decisions by themselves to pursue their individual objectives. Therefore, they may not follow the strategy optimizing the system objectives, and thus unilaterally deviate from it to maximize their own utilities.
- Due to the explosion of network size and lack of central node, the global information may be difficult to obtain, and the computational complexity may grow exponentially with the network size. In consequence, the centralized methods with global information demand and high computational complexity may be difficult to be deployed in practice.

As a result, the self-organized distributed task offloading algorithms with individual objectives, which require local information and low complexity, may be more desirable.

Furthermore, from the view of tasks, tasks can be grouped into two classes: non-splittable tasks and splittable tasks. For non-splittable tasks, they need to be processed as a whole. That is to say, a task can be either executed by its own TN or entirely offloaded to one neighboring HN, as shown in Fig. 3.1. While for splittable tasks, such as face detection, they can be divided into multiple subtasks and processed on multiple computation nodes in parallel [30], as shown in Fig. 3.1. Most works have focused on the non-splittable tasks, and to the best knowledge of ours, there are only a handful of studies considering the splittable tasks, especially for the MTMH problem [26, 31, 32]. Li et al. [31] considered that the tasks could be only processed on local TN and one HN in parallel. Liu et al. [32] simply assumed that a HN could only accommodate the task from single TN. More complex and general case was considered in [26], where tasks could be divided into multiple subtasks and offloaded to multiple HNs for parallel processing, and HNs might accommodate tasks from multiple TNs. They adopted the dynamic programming to develop a centralized heuristic algorithm to minimize the overall system delay.

In this chapter, we will investigate the task offloading problem in MTMH fog networks under the case of non-splittable tasks and splittable tasks, respectively. We propose a game theory based analytical framework and a corresponding distributed algorithm for each. To be specific, we will first introduce the general MTMH fog networks and the corresponding task offloading problem for non-splittable tasks and splittable tasks, respectively. Then, we will discuss the fundamentals of game theory and how to apply it to formulate and tackle the task offloading problems introduced above. More specifically, a Paired Offloading of Multiple Tasks (POMT) game based on potential games and a Parallel Offloading of Splittable Tasks (POST) game based on Generalized Nash equilibrium problem (GNEP) are formulated and studied thoroughly. The corresponding distributed algorithms, namely POMT and POST, are also designed. Finally, extensive simulations are conducted to demonstrate the performance of the proposed POMT and POST algorithms.

3.2 System Architecture and Analytical Models

3.2.1 Multi-Task Multi-Helper Fog Networks

Consider a general heterogeneous fog network consisting of many randomly distributed FNs with diverse computation resources and capabilities. For the sake of description and analysis, the quasi-static scenario is considered, where the time is divided into identical small time slots, such that the network condition and user distribution can be treated as static in each time slot. As shown in Fig. 3.1, at any specific time slot, the FNs can be further classified as (1) TNs, which have a task to process, (2) HNs, which have spare resources to help their neighboring TNs to process tasks, or (3) busy nodes (BNs), which are busy with processing previous tasks and thus have no spare resources. It is worth noting that TNs, HNs, and BNs are not fixed. To be specific, TNs and BNs can be HNs when they are idle and available in the following time slots, and vice versa.

Such an MTMH fog network is a quite general model and can be applied in different application scenarios. For fog-enabled robot systems, TNs can be the robots, and HNs can be the idle robots, edge nodes in the room, factory. For fog-enabled wireless communication networks, TNs can be the mobile users, and HNs can be the AP, BS in the vicinity. For fog-enabled intelligent transportation systems, TNs can be the vehicles, and HNs can be the roadside units around. For fog-enabled smart home, TNs can be the heterogeneous sensors, devices, and HNs can be the routers, local edge nodes in the house. For the MTMH fog networks, many elements need to be considered, such as task scheduling, resource allocation, mobility management, security mechanism, and so on. In this chapter, we mainly take the task scheduling as an example. More specifically, we will focus on the task scheduling problem in MTMH fog networks under two different types of tasks: non-splittable tasks and splittable tasks.

As aforementioned, tasks with strict delay requirements may not be accomplished in time on the local devices, and thus they need to be offloaded to neighboring available HNs. Task scheduling studies how to schedule tasks to HNs. Take the cases in Fig. 3.1 as examples. For non-splittable tasks, a task can be either executed by the local TN/device, or entirely offloaded to one neighboring HN. While a HN, if its spare resources and capabilities permit, can accommodate multiple tasks from different TNs. For example, as shown in Fig. 3.1a, TN-1 executes its task on local device, which may be due to (1) HN-2 does not have sufficient computation/storage resources for accommodating the task from TN-1, or (2) the communication channel between TN-1 and TN-2 is not good at this slot. TN-2 offloads its task to HN-1, because HN-1 has more computation resources and better communication channel than HN-2. HN-4 accommodates multiple tasks from TN-3 and TN-4 simultaneously as it has sufficient resources and capabilities. While for splittable tasks, a task can be divided into multiple independent subtasks and offloaded to multiple HNs for parallel processing. For example, as shown in Fig. 3.1b, TN-1 processes its task on local device, which may be due to that the neighboring BNs are busy with processing previous tasks and TN-1 is also out of the coverage of HN-1. TN-2 is in the coverage of HN-3, and thus it divides the task into two subtasks, i.e., subtask 2-1 and subtask 2-2. It processes subtask 2-1 by itself, and meanwhile offloads subtask 2-2 to HN-3 for parallel processing. TN-3 can access both HN-2 and HN-3. It divides its task into three subtasks, i.e., subtask 3-1, subtask 3-2, and subtask 3-3, and processes them on local device, HN-2, and HN-3 meanwhile. As a result, the task of TN-3 is processed on TN-3, HN-2, and HN-3 in parallel. HN-3 accommodates subtask 2-2, subtask 3-3 from TN-2, TN-3 simultaneously as it has sufficient resources and capabilities.

For the sake of analysis, denote the set of TNs (i.e., tasks) and the set of HNs by $\mathcal{N} = \{1, 2, \ldots, N\}$ and $\mathcal{K} = \{1, 2, \ldots, K\}$, respectively. For convenience, we use the task n interchangeably for the task of TN n in the following context. Then, the task offloading problem is indeed the task allocation problem between N TNs and K HNs. To be specific, the task allocation strategy can be mathematically described by the matrix $\mathbf{A} \in [0, 1]^{N\times(K+1)}$, whose (n, k)-th entry is denoted by $a_n^k \in [0, 1]$, $n \in \mathcal{N}, k \in \{0\} \cup \mathcal{K}$. a_n^k is the portion of task n processed on HN k. It is worth noting that for non-splittable tasks, $a_n^k \in \{0, 1\}$, $n \in \mathcal{N}, k \in \{0\} \cup \mathcal{K}$. Obviously, $a_n^0 + \sum_{k\in\mathcal{K}} a_n^k = 1$ should be satisfied. Notably, index 0 represents the local TN. Specifically, taking TN n as an example, a_n^0 of the task n is processed on local device and a_n^k of the task n is offloaded to HN k. Rewrite $\mathbf{A} = (\mathbf{a}_1^T, \mathbf{a}_2^T, \ldots, \mathbf{a}_N^T)^T$, where $\mathbf{a}_n = (a_n^0, a_n^1, \ldots, a_n^K)$ is the task allocation strategy of TN n.

In this chapter, we focus on the delay-minimization task offloading problem. When it comes to the task offloading, a three-phase protocol usually needs to be considered. The TNs first transmit tasks to HNs, and then after the computation of HNs, the results are transmitted back to TNs from HNs. Thus, to characterize the delay experienced by tasks, both communication delay and computation delay need to be involved. Here, it is worth noting that similar to most previous works [13, 15, 18–21, 31], the phase that HNs transmitting results back to TNs can be ignored because the size of results is negligible compared to that of tasks.

With regard to the communication aspect, we assume that all FNs are equipped with single antenna and HNs occupy orthogonal wireless channels, i.e., no interference between different HNs [11]. In this work, the channel is pre-allocated, which is beyond the scope of discussion. We also assume that the HNs serve the TNs via TDMA or frequency division multiple access (FDMA) scheme, which is consistent with most current wireless standards [23]. Thus, the TNs will share or compete for the communication resources, e.g., time frames [12] or resource blocks [26], when they transmit tasks to the same HN. Here, taking TDMA scheme as an example, the TNs are assumed to equally share the time slots. Further, we assume that a HN begins computing tasks until all subtasks offloaded to it have finished transmitting. The benefits of these assumptions are two-folds. First, it circumvents the complex transmission scheduling or resource allocation problem [10, 33], and thus the problem can be simplified. Second, the resulted transmission time can be taken as an estimate or upper bound of the actual transmission time [13]. Thus, the time of transmitting subtask from TN n to HN k can be represented as [9]

$$T_n^{k,\text{trans}} = \sum_{n \in \mathcal{N}} a_n^k \frac{z_n}{R_n^k}, \tag{3.1}$$

where z_n is the size of task n in bits. R_n^k is the data rate of transmitting task n to HN k and can be computed by the Shannon formula

$$R_n^k = B_k \log_2 \left(1 + \frac{p_n g_n^k}{\varpi_0}\right), \tag{3.2}$$

where B_k is the bandwidth occupied by HN k. p_n is the transmit power of TN n. g_n^k is the channel gain between TN n and HN k. And, ϖ_0 is the white noise power. In this work, FNs are assumed to have full channel state information (CSI).

With regard to the computation aspect, the computation time can be characterized by dividing the computation workload by the computation capability of devices, which are both in CPU cycles. Typically, computation load of each subtask is proportional to the size of subtask [30]. Thus, as many previous works, the time of computing TN n's subtask a_n^0 on local device, i.e., local computing, is given by

$$T_n^{0,\text{comp}} = a_n^0 \frac{z_n \gamma_n}{f_n}, \tag{3.3}$$

where γ_n is the processing density of task n, i.e., the CPU cycles required to process a unit bit of data, and f_n is the computation capability of TN n, i.e., the CPU clock frequency (in CPU cycles per second). Different from local computing, for subtasks offloaded to HNs, i.e., HN computing, there may be competition or sharing among tasks for the computation capability of HNs. That is because the computation capability of single HN is usually limited and cannot be exclusively used by single task. Here, it is assumed that every HN assigns its spare computation resources to all the existing tasks in proportion to their computation workloads. Again, this assumption avoids the complex computation scheduling problem and the resulted computation time can be taken as an estimate or upper bound of the actual computation time [9]. Therefore, the time of computing task n's subtask on HN k can be written as

$$T_n^{k,\text{comp}} = \sum_{n \in \mathcal{N}} a_n^k \frac{z_n \gamma_n}{f_k}, \tag{3.4}$$

where f_k is the spare computation capability of HN k.

In conclusion, for local computing, since no communication is necessary, the time required to process TN n on local device can be represented as

$$T_n^0 = a_n^0 \frac{z_n \gamma_n}{f_n}. \tag{3.5}$$

For HN computing, it involves both communication delay and computation delay. According to the communication model and the computation model introduced above, the time required to process TN n on HN k can be represented as

$$T_n^k = \sum_{n \in \mathcal{N}} a_n^k z_n \left(\frac{1}{R_n^k} + \frac{\gamma_n}{f_k} \right). \tag{3.6}$$

For ease of exposition, we define $O_n^k \triangleq z_n(1/R_n^k + \gamma_n/f_k)$, which represents the amount of time that task n contributes to the total time for offloading. Thus, $T_n^k = \sum_{n \in \mathcal{N}} a_n^k O_n^k$.

3.2.2 Non-Splittable Task and Paired Offloading

For non-splittable tasks, a task is either executed on the local device, or entirely offloaded to one neighboring HN. Thus, the MTMH problem is actually the pairing strategy between TNs and HNs for achieving the minimum processing delay of every task. For ease of presentation and analysis, we redefine the task allocation strategy of TN n as the pairing strategy $a_n = k$, where $k : a_n^k = 1$ for $k \in \{0\} \cup \mathcal{K}$, and the overall pairing profile as $\mathbf{a} = (a_1, a_2, \ldots, a_N)$. To be specific, we have $a_n > 0$, if TN n is paired with HN a_n; we have $a_n = 0$, if TN n decides to process its task on local device. Since a task is either executed on the local device, or entirely offloaded to one neighboring HN, the time required to process task n can be represented as

$$T_n(a_n, \mathbf{a}_{-n}) = \begin{cases} \dfrac{z_n \gamma_n}{f_n}, & a_n = 0 \\ \displaystyle\sum_{m \neq n} O_m^{a_m} I_{\{a_m = a_n\}} + O_n^{a_n}, & a_n > 0 \end{cases}, \tag{3.7}$$

where $I_{\{x\}}$ is an indicator function and $\mathbf{a}_{-n} = \{a_1, \ldots, a_{n-1}, a_{n+1}, \ldots, a_N\}$ is the pairing strategies of all TNs except n. Specifically, if x is true, $I_{\{x\}} = 1$; otherwise, $I_{\{x\}} = 0$.

Introduce a matrix $\mathbf{C} = (\mathbf{c}_1^T, \mathbf{c}_2^T, \ldots, \mathbf{c}_N^T)^T$ to represent the connectivity between TNs and HNs, where $\mathbf{c}_n = (c_n^1, \ldots, c_n^K)$. Specifically, if $c_n^k = 1$, then TN n can access HN k; otherwise, TN n cannot access HN k. The connectivity matrix can be determined by the distance or the signal to interference plus noise ratio (SINR). Every TN aims to minimize the time required to process its own task, i.e.,

$$\begin{aligned} &\min_{a_n} \; T_n(a_n, \mathbf{a}_{-n}), \; \forall n \in \mathcal{N} \\ &s.t. \; a_n \in \{0\} \cup \{k \in \mathcal{K} | c_n^k = 1\}, \forall n \in \mathcal{N}. \end{aligned} \tag{3.8}$$

Such a problem is a complicated combinatorial optimization problem, which is NP hard to be solved. The pairing strategies between TNs and HNs are coupled. Specifically, the time required to process task n depends on not only the pairing strategy of TN n, but also other TNs' pairing strategies, for the reason that there exists competition for the communication resources and computation capabilities of HNs among different TNs.

3.2.3 Splittable Task and Parallel Offloading

For splittable tasks, a task can be divided into multiple independent subtasks and offloaded to multiple HNs for parallel processing. Thus, the final execution time of tasks is decided by the most time-consuming subtask and the total time required to process task n can be represented as

$$T_n(\mathbf{a}_n, \mathbf{A}_{-n}) = \max_{k \in \{0\} \cup \mathcal{K}} \{T_n^k\}, \tag{3.9}$$

where $\mathbf{A}_{-n} = (\mathbf{a}_1^T, \ldots, \mathbf{a}_{n-1}^T, \mathbf{a}_{n+1}^T \ldots, \mathbf{a}_N^T)^T$ is the task allocation strategies of all TNs except n. It is worth noting that how to combine results is out of the scope of this work, and the corresponding time is ignored in this work.

In this work, every TN wants to make decisions by themselves and aims to minimize the time required to process its own task, i.e.,

$$\min_{a_n, T_n} T_n, \tag{3.10a}$$

$$\text{s.t. } a_n^0 \frac{z_n \gamma_n}{f_n} \leq T_n, \ \forall n \in \mathcal{N}, \tag{3.10b}$$

$$a_n^k \left[\sum_{n \in \mathcal{N}} a_n^k O_n^k \right] \leq a_n^k T_n, \ \forall k \in \mathcal{K}, \tag{3.10c}$$

$$a_n^0, a_n^k \geq 0, \ \forall k \in \mathcal{K}, \tag{3.10d}$$

$$a_n^0 + \sum_{k \in \mathcal{K}} a_n^k = 1, \ \forall n \in \mathcal{N}, \tag{3.10e}$$

$$a_n^k \leq c_n^k, \forall k \in \mathcal{K}. \tag{3.10f}$$

Constraint (3.10b) shows that the local computing time should be less than the total time, while constraint (3.10c) indicates that the HN computing time should also be less than the total time. Constraints (3.10d)–(3.10e) ensure that tasks are completely accomplished. Constraint (3.10f) guarantees that no TN can offload tasks to HNs, which it cannot access.

Such a problem is a complicated non-convex and non-integer optimization problem, which is also NP hard to be solved. Compared with the counterpart problem for non-splittable tasks, this problem is somehow more challenging: (1) the task allocation strategies of TNs are also coupled, (2) not only the HN selection problem but also the task division problem need to be tackled.

3.3 Theoretical Preliminaries

3.3.1 *Potential Game*

Potential games are an important branch of game theory, more specifically the non-cooperative games. Thus, before looking into the potential games, let us first briefly introduce the fundamentals of non-cooperative games.

Game theory can be viewed as a branch of applied mathematics as well as of applied sciences. It has been used in the social sciences, most notably in economics, but has also penetrated into a variety of other disciplines such as political science, biology, computer science, philosophy, and, recently, wireless and communication networks. Non-cooperative game theory is one of the most important branches of game theory, focusing on the study and analysis of competitive decision-making involving several players. It provides an analytical framework suited for characterizing the interactions and decision-making process involving several players with partially or totally conflicting interests over the outcome of a decision process which is affected by their actions [34].

A non-cooperative game can be defined by the strategic or normal form which has three components: the set of players, their strategies, and the payoffs/costs or utilities. More formally, it can be defined as follows [34]:

Definition 3.1 A non-cooperative game in strategic or normal form is a triplet $G = (\mathcal{N}, (\mathcal{S}_i)_{i\in\mathcal{N}}, (u_i)_{i\in\mathcal{N}})$, where

- $\mathcal{N}$ is a finite set of players, i.e., $\mathcal{N} = \{1, \ldots, N\}$.
- $\mathcal{S}_i$ is the set of available strategies for player i.
- $u_i : \mathcal{S} \to \mathbb{R}$ is the utility (payoff/cost) function for player i, with $\mathcal{S} = \mathcal{S}_1 \times \cdots \times \mathcal{S}_i \times \cdots \times \mathcal{S}_N$ (Cartesian product of the strategy sets).

Given the definition of a strategic game, for any player i, every element $s_i \in \mathcal{S}_i$ is the strategy of player i, $\mathbf{s}_{-i} = [s_j]_{j\in\mathcal{N}, j\neq i}$ denotes the vector of strategies of all players except i, and $\mathbf{s} = (s_i, \mathbf{s}_{-i}) \in \mathcal{S}$ is called as the strategy profile. If the sets of players' strategies $\mathcal{S}_i$ are finite for all $i \in \mathcal{N}$, the game is called finite; otherwise, it is called infinite. And if each player $i \in \mathcal{N}$ selects a strategy $s_i \in \mathcal{S}_i$ in a deterministic manner, i.e., with probability 1, then this strategy is known as a pure strategy.

For game theory, one important solution concept is the Nash equilibrium (NE). An NE is a state where no player can improve its utility or reduce its payoff/cost by changing its strategy, if the other players maintain their current strategies. Formally, when dealing with pure strategies, i.e., the choices of players are deterministic, the NE is defined as follows [34]:

Definition 3.2 A pure-strategy NE of the non-cooperative game $G = (\mathcal{N}, (\mathcal{S}_i)_{i\in\mathcal{N}}, (u_i)_{i\in\mathcal{N}})$ is a strategy profile $\mathbf{s}^* \in \mathcal{S}$ such that

$$u_i(s_i^*, \mathbf{s}_{-i}^*) \geq u_i(s_i, \mathbf{s}_{-i}^*), \ \forall i \in \mathcal{N}, \ \forall s_i \in \mathcal{S}_i. \tag{3.11}$$

If $u_i(s_i^*, \mathbf{s}_{-i}^*) > u_i(s_i, \mathbf{s}_{-i}^*), \ \forall i \in \mathcal{N}, \ \forall s_i \in \mathcal{S}_i, \ s_i \neq s_i^*, \ \forall i \in \mathcal{N}$, the NE is said to be strict. In other words, a strategy profile is a pure-strategy NE if no player has an incentive to unilaterally deviate to another strategy, given that other players' strategies remain fixed.

When studying the NE of a game, the key points of interest are existence, multiplicity, and efficiency. It has been proven that a non-cooperative game can admit zero, one, or multiple NEs (may be the mixed NE), but the pure-strategy NE is not guaranteed to exist. Efficiency says that an NE is not necessarily the best outcome, from the perspective of payoff.

When solving the non-cooperative games, the concept of the best response function is useful, which is defined as follows [34]:

Definition 3.3 The best response function $b_i(\mathbf{s}_{-i})$ of a player i to the profile of strategies $\mathbf{s}_{-i}$ is a set of strategies for that player such that

$$b_i(\mathbf{s}_{-i}) = \{s_i \in \mathcal{S}_i | u_i(s_i, \mathbf{s}_{-i}) \geq u_i(s_i', \mathbf{s}_{-i}), \ \forall s_i' \in \mathcal{S}_i\}. \tag{3.12}$$

Hence, for a player i, when the strategies of the other players are fixed as $\mathbf{s}_{-i}$, any strategy in $b_i(\mathbf{s}_{-i})$ is at least as good as every other available strategy in $\mathcal{S}_i$. The best response function is set-valued as it associates, for a player i, a set of strategies with any strategy profile $\mathbf{s}_{-i}$ of the other players. Every element of the best response function $b_i(\mathbf{s}_{-i})$ is a best response of player i to $\mathbf{s}_{-i}$. In other words, the best response of a player i implies that, if each of the other players adheres to $\mathbf{s}_{-i}$, then player i cannot do better than to choose a member of $b_i(\mathbf{s}_{-i})$.

The concept of the best response function leads to an alternative characterization of a pure-strategy NE [34].

Theorem 3.1 *A strategy profile $\mathbf{s}^* \in \mathcal{S}$ is a pure-strategy NE of a non-cooperative game if and only if every player's strategy is a best response to the other players' strategies, i.e.,*

$$s_i^* \in b_i(\mathbf{s}_{-i}^*), \ \forall i \in \mathcal{N}. \tag{3.13}$$

After briefly introducing the fundamentals of game theory, let us look into potential games. Among different non-cooperative games, potential games are an important branch. What make potential games attractive are their useful properties concerning the existence and attainability of their NE. Potential game has four different types: ordinal potential games, weighted potential games, exact (cardinal) potential games, and generalized ordinal potential games. Due to the space limit, we will only introduce the ordinal potential games, weighted potential games, and exact (cardinal) potential games. For more details about potential games, readers can refer to [35].

Definition 3.4 The game G is an exact (cardinal) potential game if and only if a potential function $\Phi(\mathbf{s}) : \mathscr{S} \rightarrow \mathscr{R}$ exists such that $\forall i \in \mathscr{N}$,

$$u_i(s_i', \mathbf{s}_{-i}) - u_i(s_i, \mathbf{s}_{-i}) = \Phi(s_i', \mathbf{s}_{-i}) - \Phi(s_i, \mathbf{s}_{-i}), \forall s_i, s_i', \ \forall \mathbf{s}_{-i} \in \mathscr{S}_{-i}. \quad (3.14)$$

In exact potential games, the change in a single player's utility due to his/her own strategy deviation results in exactly the same amount of change in the potential function.

Definition 3.5 The game G is a weighted potential game if and only if a potential function $\Phi(\mathbf{s}) : \mathscr{S} \rightarrow \mathscr{R}$ exists such that $\forall i \in \mathscr{N}$,

$$u_i(s_i', \mathbf{s}_{-i}) - u_i(s_i, \mathbf{s}_{-i}) = w_i(\Phi(s_i', \mathbf{s}_{-i}) - \Phi(s_i, \mathbf{s}_{-i})), \forall s_i, s_i', \ \forall \mathbf{s}_{-i} \in \mathscr{S}_{-i}, \quad (3.15)$$

where $(w_i)_{i \in \mathscr{N}}$ is a vector of positive numbers, known as the weights. In weighted potential games, a player's change in payoff due to his/her unilateral strategy deviation is equal to the change in the potential function but scaled by a weight factor. Clearly, all exact potential games are weighted potential games with all players having identical weights of one.

Definition 3.6 The game G is an ordinal potential game if and only if a potential function $\Phi(\mathbf{s}) : \mathscr{S} \rightarrow \mathscr{R}$ exists such that $\forall i \in \mathscr{N}$,

$$u_i(s_i', \mathbf{s}_{-i}) - u_i(s_i, \mathbf{s}_{-i}) > 0 \Leftrightarrow \Phi(s_i', \mathbf{s}_{-i}) - \Phi(s_i, \mathbf{s}_{-i}), \forall s_i, s_i', \ \forall \mathbf{s}_{-i} \in \mathscr{S}_{-i}. \quad (3.16)$$

In ordinal potential games, if player i gains a better (worse) utility from switching his/her strategy, this should lead to an increase (decline) in the potential function, and vice versa. Clearly, exact (cardinal) potential games and weighted potential games are both ordinal potential games.

As aforementioned, potential games have good properties concerning the existence and attainability of their NE, which is concluded as [35]:

Theorem 3.2 *Every finite (ordinal) potential game admits at least one pure-strategy NE.*

Theorem 3.3 *For finite exact (cardinal) potential games, every improvement path is finite and every sequence of better and best responses converges to an NE, regardless of its starting point. This is known as the finite improvement property.*

3.3.2 Generalized Nash Equilibrium Problem

The GNEP is an important model that has its roots in the economic sciences but is being fruitfully used in many different fields [36]. As the GNEP lies at the intersection of many different disciplines, it has a number of different names in the literature including pseudo-game, social equilibrium problem, equilibrium programming, coupled constraint equilibrium problem, and abstract economy.

Formally, the GNEP also consists of players, strategies, and utility/payoff/cost functions. But different from common non-cooperative games, each player's strategy must belong to a set $\mathbf{S}_i(\mathbf{s}_{-i})$ that depends on the rival players' strategies and that we call the feasible set or strategy space of player i. The aim of player i, given the other players' strategies $\mathbf{s}_{-i}$, is to choose a strategy s_i that solves the maximization problem (for utility functions)

$$\text{maximize}_{s_i}\ u_i(s_i, \mathbf{s}_{-i}) \text{ subject to } s_i \in \mathbf{S}_i(\mathbf{s}_{-i}). \tag{3.17}$$

The solution to the problem (3.17) can be taken as the best response function introduced above, and thus the NE of GNEP can also be defined as Theorem 3.1. It is worth noting that if the feasible set $\mathbf{S}_i(\mathbf{s}_{-i})$ do not depend on the rival players' strategies, i.e., $\mathbf{S}_i(\mathbf{s}_{-i}) = \mathscr{S}_i$, the GNEP reduces to the standard Nash equilibrium problem.

3.3.3 Matching Theory

When applying the conventional non-cooperative games to solve practical problems, it may come into some problems. First, classical game-theoretic algorithms, such as the best response mechanism, require the knowledge of other players' actions, thus limiting their distributed implementations. Second, most game-theoretic solutions, such as the NE, investigate the one-sided (or unilateral) stability notions in which the equilibrium deviations are evaluated unilaterally. Such unilateral deviations may not be practical when investigating the assignment problems between distinct sets of players. Last, but not least, the tractability of equilibrium in the game-theoretic methods requires having certain form of structure in the objective function, which cannot be satisfied in some practical applications.

Matching theory has emerged as a promising technique which can overcome some limitations of the conventional non-cooperative games. Matching theory is a Nobel Prize winning framework that provides mathematically tractable solutions for the combinatorial problem of matching players in two distinct sets, depending on the individual information and preference of each player. The advantages of matching theory for network management include: (1) suitable models for characterizing interactions between the heterogeneous nodes, each of which has its own type, objective, and information, (2) the ability to define the general "preferences" that can handle the heterogeneous and complex considerations related to the quality-of-service (QoS) requirements, (3) suitable solutions, in terms of stability and optimality, that can accurately reflect different system objectives, and (4) efficient algorithmic implementations that are inherently self-organizing [37].

The simplest and oldest version of the matching problem is the marriage problem. The problem involves a set of M of men and a set W of women. Each $m \in \mathcal{M}$ has a strict preference ordering over the elements of W and each $w \in \mathcal{W}$ has a strict preference ordering over the men. The preference ordering of agent i will be denoted $\succ_i$ and $x \succ_i y$ means that agent i ranks x above y. A matching is an assignment of men to women such that each man is assigned to at most one woman and vice versa. Mathematically, this matching can be defined as:

Definition 3.7 A matching for marriage problem is $\mu : \mathcal{M} \cup \mathcal{W} \Rightarrow 2^{\mathcal{M} \cup \mathcal{W}}$ such that:

- $\mu(m) \subseteq \mathcal{W}$ such that $|\mu(m)| \leq 1$ for all $m \in \mathcal{M}$.
- $\mu(w) \subseteq \mathcal{M}$ such that $|\mu(w)| \leq 1$ for all $w \in \mathcal{W}$.
- $m = \mu(w)$ if and only if $\mu(w) = m$ for all $m \in \mathcal{M}$ and $w \in \mathcal{W}$.

In matching theory, stable is an important concept. To proceed, we need to first introduce the blocking pair.

Definition 3.8 The pair $(m, w^{'})$ is called a blocking pair if

- m is matched to w.
- $m^{'}$ is matched to $w^{'}$.
- $w^{'} \succ_m w$ and $m \succ_{w^{'}} m^{'}$.

Definition 3.9 The matching μ is called stable if it has no blocking pairs.

The marriage problem is the most basic matching problem, which belongs to the One-to-one matching. Matching problems cover a wide range of problems and can be classified in different ways [37]. One way is as follows, which considers the capacity/quota allowed for each agent:

- One-to-one matching: It means each member of one set can be matched to at most one player from the opposite set. Examples include marriage problem, forming roommate pairs, and so on.
- Many-to-one matching: It means each agent of one set can be matched to more than one member from the opposite set up to the capacity, while agents from the opposite side can only be matched to one agent at most. Examples are like allocating residents to hospitals, assigning school leavers to universities.
- Many-to-many matching: It means agents from both matching sets can be matched to more than one agent up to their capacities. Examples include creating partnerships in P2P networks and assigning workers to firms problem.

Another popular classification way is:

- Bipartite matching problems with two-sided preferences: Here the participating agents can be partitioned into two disjoint sets, and each member of one set ranks a subset of the members in the other set in order of preference. Example applications include assigning junior doctors to hospitals, pupils to schools, and school leavers to universities.
- Bipartite matching problems with one-sided preferences: Again the participating agents can be partitioned into two disjoint sets, but this time only one set of players rank the subsets of the members in the other set in order of preferences. Example applications include campus housing allocation, DVD rental markets and assigning reviewers to conference papers.
- Non-bipartite matching problems with preferences: Here, all the participating agents form a single homogeneous set, and each agent ranks a subset of the others in order of preferences. Example applications include forming pairs of agents for chess tournaments, finding kidney exchanges involving incompatible (patient, donor) pairs, and creating partnerships in P2P networks.

3.4 Game Formulation and Algorithm Design

3.4.1 Paired Offloading of Multiple Tasks

To analyze the task offloading for non-splittable tasks, we can formulate the interactions among TNs as a potential game. As mentioned above, every TN wants to minimize the time required to process its own task, while the pairing strategies among TNs are coupled. To be specific, if too many TNs choose to offload tasks to the same HN, they may incur serious execution time increase to each other (as shown in formula (3.7)). Thus, we can model the interactions among TNs as a non-cooperative game. But, the way to define the individual execution time of tasks as the cost of TNs usually greatly damages the system performance. Therefore, to balance the TNs' selfishness and the system performance, the local altruistic behavior among TNs is introduced when constructing the cost function [38, 39].

Formally, we define our POMT game [9] as $G_1 = (\mathcal{N}, (\mathcal{S}_n)_{n\in\mathcal{N}}, (u_n)_{n\in\mathcal{N}})$, where $\mathcal{S}_n : \{0\} \cup \{k \in \mathcal{K} | c_n^k = 1\}$ is the pairing strategy space of TN n and u_n is the cost function which can be expressed as

$$u_n(a_n, \mathbf{a}_{-n}) = \begin{cases} \dfrac{z_n \gamma_n}{f_n}, & a_n = 0 \\ \sum\limits_{m \neq n} \left(O_m^{a_m} + O_n^{a_n}\right) I_{\{a_m = a_n\}} + O_n^{a_n}, & a_n > 0 \end{cases}. \tag{3.18}$$

Notably, different from (3.7), we include the influence to others incurred by TN n (the first $O_n^{a_n}$ term in the second line of (3.18)) when constructing the cost function, i.e., local altruistic behaviors.

We next study the existence of pure-strategy NE for POMT game G_1, which is concluded in the following Proposition 3.1.

Proposition 3.1 *The POMT game G_1 possesses at least one pure-strategy NE and guarantees the finite improvement property.*

Proof We first prove that the POMT game G_1 is an exact (cardinal) potential game with potential function

$$\begin{aligned}\Phi(a_n, \mathbf{a}_{-n}) =& \frac{1}{2}\sum_{n\in\mathcal{N}}\sum_{m\neq n}(O_m^{a_m}+O_n^{a_n})I_{\{a_m=a_n\}}I_{\{a_n>0\}} \\ &+\sum_{n\in\mathcal{N}} O_n^{a_n} I_{\{a_n>0\}} + \sum_{n\in\mathcal{N}}\frac{z_n\gamma_n}{f_n}I_{\{a_n=0\}},\end{aligned} \tag{3.19}$$

such that

$$\begin{aligned}\Phi(a_n', \mathbf{a}_{-n}) - \Phi(a_n, \mathbf{a}_{-n}) = c_n(a_n', \mathbf{a}_{-n}) - c_n(a_n, \mathbf{a}_{-n}),\\ \forall a_n,\ a_n' \in \mathcal{A}_n,\ \mathbf{a}_{-n} \in \prod_{m\neq n}\mathcal{A}_m.\end{aligned} \tag{3.20}$$

Case 1: $a_n > 0$, $a_n' > 0$

$$\begin{aligned}&\Phi(a_n', \mathbf{a}_{-n}) - \Phi(a_n, \mathbf{a}_{-n})\\ &= \frac{1}{2}\sum_{m\neq n}(O_m^{a_m}+O_n^{a_n'})I_{\{a_m=a_n'\}} + \frac{1}{2}\sum_{m\neq n}(O_n^{a_n'}+O_m^{a_m})I_{\{a_n'=a_m\}}\\ &\quad -\frac{1}{2}\sum_{m\neq n}(O_m^{a_m}+O_n^{a_n})I_{\{a_m=a_n\}} - \frac{1}{2}\sum_{m\neq n}(O_n^{a_n}+O_m^{a_m})I_{\{a_n=a_m\}}\\ &\quad + O_n^{a_n'} - O_n^{a_n}\\ &= \sum_{m\neq n}(O_m^{a_m}+O_n^{a_n'})I_{\{a_m=a_n'\}} + O_n^{a_n'}\\ &\quad - \sum_{m\neq n}(O_m^{a_m}+O_n^{a_n})I_{\{a_m=a_n\}} - O_n^{a_n}\\ &= c_n(a_n', \mathbf{a}_{-n}) - c_n(a_n, \mathbf{a}_{-n})\end{aligned} \tag{3.21}$$

Case 2: $a_n = 0$, $a'_n > 0$

$$\begin{aligned}
&\Phi(a'_n, \mathbf{a}_{-n}) - \Phi(a_n, \mathbf{a}_{-n}) \\
&= \frac{1}{2}\sum_{m \neq n}(O_m^{a_m} + O_n^{a'_n}) I_{\{a_m = a'_n\}} + \frac{1}{2}\sum_{m \neq n}(O_n^{a'_n} + O_m^{a_m}) I_{\{a'_n = a_m\}} \\
&+ O_n^{a'_n} - \frac{z_n \gamma_n}{f_n} \\
&= \sum_{m \neq n}(O_m^{a_m} + O_n^{a'_n}) I_{\{a_m = a'_n\}} + O_n^{a'_n} - \frac{z_n \gamma_n}{f_n} \\
&= c_n(a'_n, \mathbf{a}_{-n}) - c_n(a_n, \mathbf{a}_{-n})
\end{aligned} \tag{3.22}$$

Case 3: $a_n > 0$, $a'_n = 0$

Similar to case 2, it is straightforward to show that $\Phi(a'_n, \mathbf{a}_{-n}) - \Phi(a_n, \mathbf{a}_{-n}) = c_n(a'_n, \mathbf{a}_{-n}) - c_n(a_n, \mathbf{a}_{-n})$.

In conclusion, the POMT game G_1 is an exact potential game with the potential function as given in (3.19). For finite potential games, there exists at least one pure-strategy NE. Furthermore, every sequence of better and best responses converges to an NE, regardless of its starting point, i.e., the finite improvement property [40].

As stated in Theorem 3.2, any asynchronous better or best response update process is guaranteed to reach a pure-strategy NE within a finite number of iterations, regardless of its starting point and the order of updates. By employing such a property, we can design a distributed computation offloading algorithm called POMT to obtain an NE for the POMT game, as shown in [15].

Consider a slotted time structure, and every offloading time slot is further divided into multiple decision slots. To begin with, every TN chooses local computing, i.e., $a_n(0) = 0$ (line 2). Then, at every decision slot t:

1. *TNs broadcasting parameters to HNs*: the TNs simultaneously broadcast the parameters message to the achievable HNs, i.e., the HNs within whose coverage they are (line 5). The message informs the HNs of the task information, i.e., the task size and processing density, the transmit power, and so on.
2. *HNs broadcasting measurements to TNs*: the HNs first estimate the channel conditions and calculate the measurements of total processing time, i.e., $\sum_{n \in \mathcal{N}} O_n^k I_{\{a_n = k\}}$. Then, they broadcast the measurements back to the TNs (line 6).
3. *TNs computing the best response functions and contending for the update opportunity*: based on the measurements from the HNs, the TNs can compute the best response functions, i.e., $b_n(\mathbf{a}_{-n}(t))$ (line 7), and then decide whether

to contend for the update opportunity (line 8–9). To be specific, if $a_n(t) \notin b_n(\mathbf{a}_{-n}(t))$, i.e., TN n can further reduce its cost by choosing $a_n \in b_n(\mathbf{a}_{-n}(t))$, TN n sends a request to update (RTU) message to the corresponding HN $a_n \in b_n(\mathbf{a}_{-n}(t))$ to contend for the pairing strategy update opportunity.

4. *HNs deciding the updated TN and broadcasting the update message*: once receiving the RTU messages, the HNs collaborate with each other to decide the TN which wins the update opportunity. For example, this can be realized by the following way: a master HN is in charge of gathering all RTU messages from other HNs and then randomly decides the TN which wins the update opportunity. After deciding the updated TN, the HNs broadcast the update-permission (UP) message to the TNs to inform them that which TN is permitted to update its pairing strategy.
5. *TNs updating the pairing strategies*: If TN n is permitted to update its pairing strategy, then it updates its pairing strategy as $a_n(t+1) \in b_n(\mathbf{a}_{-n}(t))$ at the next decision slot (line 11). Otherwise, it maintains the current pairing strategy at the next decision slot (line 13).

If no TNs want to update the current pairing strategy, i.e., no RTU messages are sent to the HNs, the HNs will broadcast the END message to all TNs and the algorithm terminates (line 18). The whole procedure is shown as Algorithm 1.

Algorithm 1 POMT algorithm

1: **initialization**:
2: every TN n chooses the pairing strategy $a_n(0) = 0$.
3: **end initialization**
4: **repeat** for every TN n and every decision slot in parallel:
5: broadcast the parameters message to the achievable HNs.
6: receive the measurements of $\sum_{n \in \mathcal{N}} O_n^k I_{\{a_n=k\}}$ from the achievable HNs.
7: compute the best response function $b_n(\mathbf{a}_{-n}(t))$.
8: **if** $a_n(t) \notin b_n(\mathbf{a}_{-n}(t))$ **then**
9:     send RTU message to the corresponding HN for contending for the pairing strategy update opportunity.
10:     **if** not permitted to update the pairing strategy **then**
11:         update the pairing strategy $a_n(t+1) \in b_n(\mathbf{a}_{-n}(t))$ for the next decision slot.
12:     **else**
13:         maintain the current pairing strategy $a_n(t+1) = a_n(t)$ for the next decision slot.
14:     **end if**
15: **else**
16:     maintain the current pairing strategy $a_n(t+1) = a_n(t)$ for next slot.
17: **end if**
18: **until** END message is received from the HNs.

3.4.2 Parallel Offloading of Splittable Tasks

When it comes to splittable tasks, potential game theory and finite improvement proper are no longer suitable to analyze the task offloading, because the strategy sets of players are infinite. Therefore, we resort to the GNEP theory and formulate the task offloading for splittable tasks as a GNEP [41]. Formally, we define the corresponding POST game for splittable tasks as $G_2 = \{\mathcal{N}, \{\mathcal{S}_n\}_{n\in\mathcal{N}}, \{T_n\}_{n\in\mathcal{N}}\}$, where $\mathcal{S}_n = \left\{\mathbf{a}_n \in [0,1]^{1\times(K+1)} \mid a_n^0,\ a_n^k \geq 0,\ a_n^0 + \sum_{k\in\mathcal{K}} a_n^k = 1, a_n^k \leq c_n^k,\ a_n^0 \frac{z_n\gamma_n}{f_n} \leq T_n(\mathbf{a}_n, \mathbf{A}_{-n}),\ a_n^k \left[\sum_{n\in\mathcal{N}} a_n^k O_n^k\right] \leq a_n^k T_n,\ \forall k \in \mathcal{K}\right\}$ is the task allocation strategy space of TN n and $T_n = T_n(\mathbf{a}_n, \mathbf{A}_{-n})$ is the cost of TN n. Notably, this is a GNEP because the strategy set of a player depends on the rival players' strategies.

For the POST game, the existence of NE is not straightforward for the reason that the optimization problem (3.10) is non-convex (see contents on the joint convex GNEP, short for joint convex generalized Nash equilibrium problem, in [36]). To understand this point, let us see the constraint (3.10c). For $a_n^k = 0$, this constraint is automatically fulfilled, and thus vanishes. Such a constraint is called vanishing constraint, which is non-convex [42].

As stated above, although the existence of NE cannot be guaranteed by the joint convex GNEP theory, it cannot be claimed that the POST game does not possess an NE. To further prove the existence of NE for POST game, we first study the structural properties of solutions to problem (3.10), which are summarized in Propositions 3.2–3.4.

Proposition 3.2 *For the optimal solution* $(\mathbf{a}_n^*, T_n^*)$ *to the problem (3.10), it satisfies*

$$a_n^{0*} \frac{z_n\gamma_n}{f_n} = T_n^*, \tag{3.23}$$

and

$$\sum_{m\neq n} a_m^k O_m^k + a_n^{k*} O_n^k = T_n^*,\ \forall k:\ a_n^{k*} > 0. \tag{3.24}$$

Proof Define $\mathcal{A}_n := \{k \in \mathcal{K} | a_n^k > 0\}$, and choose $\widehat{\mathbf{a}}_n$ as the solution of

$$\begin{aligned} \sum_{m\neq n} a_m^k O_m^k + \widehat{a}_n^k O_n^k &= \widehat{a}_n^0 \frac{z_n\gamma_n}{f_n},\ k \in \mathcal{A}_n, \\ \widehat{a}_n^0 + \sum_{k\in\mathcal{A}_n} \widehat{a}_n^k &= 1. \end{aligned} \tag{3.25}$$

Therefore,

$$
\begin{aligned}
&\sum_{m\neq n} a_m^k O_m^k + \widehat{a}_n^k O_n^k = \left(1 - \sum_{k\in\mathscr{A}_n} \widehat{a}_n^k\right)\frac{z_n\gamma_n}{f_n} \\
\Rightarrow &\left(O_n^k + \frac{z_n\gamma_n}{f_n}\right)\widehat{a}_n^k + \frac{z_n\gamma_n}{f_n}\sum_{l\in\mathscr{A}_n\backslash k}\widehat{a}_n^l = \frac{z_n\gamma_n}{f_n} - \sum_{m\neq n} a_m^k O_m^k.
\end{aligned}
\tag{3.26}
$$

Thus, all $\widehat{a}_n^k$ for $k \in \mathscr{A}_n$ are given by the solution of the $L \times L$ linear equations system

$$
\begin{aligned}
&\begin{bmatrix}
O_n^1 + \frac{z_n\gamma_n}{f_n} & \frac{z_n\gamma_n}{f_n} & \cdots & \frac{z_n\gamma_n}{f_n} \\
\frac{z_n\gamma_n}{f_n} & O_n^2 + \frac{z_n\gamma_n}{f_n} & \cdots & \frac{z_n\gamma_n}{f_n} \\
\vdots & & \ddots & \\
\frac{z_n\gamma_n}{f_n} & & \cdots & O_n^L + \frac{z_n\gamma_n}{f_n}
\end{bmatrix}
\begin{bmatrix}
\widehat{a}_n^1 \\ \widehat{a}_n^2 \\ \vdots \\ \widehat{a}_n^L
\end{bmatrix} \\
&=\begin{bmatrix}
\frac{z_n\gamma_n}{f_n} - \sum_{m\neq n} a_m^1 O_m^1 \\
\frac{z_n\gamma_n}{f_n} - \sum_{m\neq n} a_m^2 O_m^2 \\
\vdots \\
\frac{z_n\gamma_n}{f_n} - \sum_{m\neq n} a_m^L O_m^L
\end{bmatrix},
\end{aligned}
\tag{3.27}
$$

where $L = |\mathscr{A}_n|$.

Rewrite (3.27) as

$$
\left(\underbrace{\operatorname{diag}\left(O_n^1, O_n^2, \ldots, O_n^L\right)}_{=:\boldsymbol{D}} + \boldsymbol{e}\cdot\underbrace{\frac{z_n\gamma_n}{f_n}\boldsymbol{e}^T}_{=:\boldsymbol{v}^T}\right)\widehat{\boldsymbol{a}}_n = \boldsymbol{b},
\tag{3.28}
$$

where $\boldsymbol{e}$ is a L-dimensional column vector of all ones.

Since $\boldsymbol{D}$ is nonsingular and $\boldsymbol{v}^T\boldsymbol{D}^{-1}\boldsymbol{e} \neq -1$, we can calculate $\widehat{\boldsymbol{a}}_n$ using the Sherman–Morrison formula, which is given in (3.29).

$$
\begin{aligned}
\widehat{\boldsymbol{a}}_n &= \left(\boldsymbol{D}^{-1} - \frac{\boldsymbol{D}^{-1}\boldsymbol{e}\boldsymbol{v}^T\boldsymbol{D}^{-1}}{1+\boldsymbol{v}^T\boldsymbol{D}^{-1}\boldsymbol{e}}\right)\boldsymbol{b} \\
&= \boldsymbol{D}^{-1}\boldsymbol{b} - \frac{\boldsymbol{D}^{-1}\boldsymbol{e}\boldsymbol{v}^T\boldsymbol{D}^{-1}\boldsymbol{b}}{1+\boldsymbol{v}^T\boldsymbol{D}^{-1}\boldsymbol{e}}.
\end{aligned}
\tag{3.29}
$$

$$
\boldsymbol{D}^{-1}\boldsymbol{b} = \begin{bmatrix} \frac{1}{O_n^1}\left(\frac{z_n\gamma_n}{f_n} - \sum\limits_{m\neq n} a_m^1 O_m^1\right) \\ \frac{1}{O_n^2}\left(\frac{z_n\gamma_n}{f_n} - \sum\limits_{m\neq n} a_m^2 O_m^2\right) \\ \vdots \\ \frac{1}{O_n^L}\left(\frac{z_n\gamma_n}{f_n} - \sum\limits_{m\neq n} a_m^L O_m^L\right) \end{bmatrix}. \tag{3.30}
$$

$$
\boldsymbol{D}^{-1}\boldsymbol{e}\boldsymbol{v}^T\boldsymbol{D}^{-1}\boldsymbol{b} = \begin{bmatrix} \frac{1}{O_n^1}\frac{z_n\gamma_n}{f_n}\sum\limits_{l\in\mathscr{A}_n}\frac{1}{O_n^l}\left(\frac{z_n\gamma_n}{f_n} - \sum\limits_{m\neq n} a_m^l O_m^l\right) \\ \frac{1}{O_n^2}\frac{z_n\gamma_n}{f_n}\sum\limits_{l\in\mathscr{A}_n}\frac{1}{O_n^l}\left(\frac{z_n\gamma_n}{f_n} - \sum\limits_{m\neq n} a_m^l O_m^l\right) \\ \vdots \\ \frac{1}{O_n^L}\frac{z_n\gamma_n}{f_n}\sum\limits_{l\in\mathscr{A}_n}\frac{1}{O_n^l}\left(\frac{z_n\gamma_n}{f_n} - \sum\limits_{m\neq n} a_m^l O_m^l\right) \end{bmatrix}. \tag{3.31}
$$

$$
\boldsymbol{v}^T\boldsymbol{D}^{-1}\boldsymbol{e} = \frac{z_n\gamma_n}{f_n}\sum_{l\in\mathscr{A}_n}\frac{1}{O_n^l}. \tag{3.32}
$$

Therefore, for every $k \in \mathscr{A}_n$, we have

$$
\begin{aligned} \widehat{a}_n^k = & \frac{1}{O_n^k}\left(\frac{z_n\gamma_n}{f_n} - \sum_{m\neq n} a_m^k O_m^k\right) \\ & - \frac{\frac{1}{O_n^k}\frac{z_n\gamma_n}{f_n}\sum\limits_{l\in\mathscr{A}_n}\frac{1}{O_n^l}\left(\frac{z_n\gamma_n}{f_n} - \sum_{m\neq n} a_m^l O_m^l\right)}{1+\frac{z_n\gamma_n}{f_n}\sum_{l\in\mathscr{A}_n}\frac{1}{O_n^l}} > 0. \end{aligned} \tag{3.33}
$$

Then,

$$
\widehat{a}_n^0 = 1 - \sum_{l\in\mathscr{A}_n}\widehat{a}_n^l = \frac{1+\sum_{l\in\mathscr{A}_n}\frac{\sum_{m\neq n} a_m^l O_m^l}{O_n^l}}{1+\frac{z_n\gamma_n}{f_n}\sum_{l\in\mathscr{A}_n}\frac{1}{O_n^l}} > 0. \tag{3.34}
$$

Set

$$
\widehat{T}_n = \widehat{a}_n^0\frac{z_n\gamma_n}{f_n} = \frac{1+\sum\limits_{l\in\mathscr{A}_n}\frac{\sum_{m\neq n} a_m^l O_m^l}{O_n^l}}{1+\frac{z_n\gamma_n}{f_n}\sum_{l\in\mathscr{A}_n}\frac{1}{O_n^l}}\frac{z_n\gamma_n}{f_n}. \tag{3.35}
$$

Obviously, $(\widehat{\mathbf{a}}_n, \widehat{T}_n)$ is a feasible solution to problem (3.10).

Assume for contradiction $a_n^{0*}\frac{z_n\gamma_n}{f_n} < T_n^*$. Then we have

$$
\begin{aligned}
\widehat{T}_n &= \frac{1+\sum_{l\in\mathscr{A}_n}\frac{\sum_{m\neq n} a_m^l O_m^l}{O_n^l}}{1+\frac{z_n\gamma_n}{f_n}\sum_{l\in\mathscr{A}_n}\frac{1}{O_n^l}}\frac{z_n\gamma_n}{f_n} \leq \frac{1+\sum_{l\in\mathscr{A}_n}\frac{T_n^*-a_n^{l*}O_n^l}{O_n^l}}{1+\frac{z_n\gamma_n}{f_n}\sum_{l\in\mathscr{A}_n}\frac{1}{O_n^l}}\frac{z_n\gamma_n}{f_n} \\
&= \frac{\left(1-\sum_{l\in\mathscr{A}_n} a_n^{l*}\right)\frac{z_n\gamma_n}{f_n}+T_n^*\frac{z_n\gamma_n}{f_n}\sum_{l\in\mathscr{A}_n}\frac{1}{O_n^l}}{1+\frac{z_n\gamma_n}{f_n}\sum_{l\in\mathscr{A}_n}\frac{1}{O_n^l}} \\
&= \frac{a_n^{0*}\frac{z_n\gamma_n}{f_n}+T_n^*\frac{z_n\gamma_n}{f_n}\sum_{l\in\mathscr{A}_n}\frac{1}{O_n^l}}{1+\frac{z_n\gamma_n}{f_n}\sum_{l\in\mathscr{A}_n}\frac{1}{O_n^l}} \\
&< \frac{T_n^*+T_n^*\frac{z_n\gamma_n}{f_n}\sum_{l\in\mathscr{A}_n}\frac{1}{O_n^l}}{1+\frac{z_n\gamma_n}{f_n}\sum_{l\in\mathscr{A}_n}\frac{1}{O_n^l}} \\
&= T_n^*,
\end{aligned}
\tag{3.36}
$$

a contradiction to the optimality of $(\mathbf{a}_n^*, T_n^*)$. Thus, for all optimal solutions, we have $a_n^{0*}\frac{z_n\gamma_n}{f_n} = T_n^*$.

Assume for contradiction $\sum\limits_{m\neq n} a_m^k O_m^k + a_n^{k*} O_n^k < T_n^*$. Then we have

$$
\begin{aligned}
\widehat{T}_n &= \frac{1+\sum_{l\in\mathscr{A}_n}\frac{\sum_{m\neq n} a_m^l O_m^l}{O_n^l}}{1+\frac{z_n\gamma_n}{f_n}\sum_{l\in\mathscr{A}_n}\frac{1}{O_n^l}}\frac{z_n\gamma_n}{f_n} < \frac{1+\sum_{l\in\mathscr{A}_n}\frac{T_n^*-a_n^{l*}O_n^l}{O_n^l}}{1+\frac{z_n\gamma_n}{f_n}\sum_{l\in\mathscr{A}_n}\frac{1}{O_n^l}}\frac{z_n\gamma_n}{f_n} \\
&= \frac{\left(1-\sum_{l\in\mathscr{A}_n} a_n^{l*}\right)\frac{z_n\gamma_n}{f_n}+T_n^*\frac{z_n\gamma_n}{f_n}\sum_{l\in\mathscr{A}_n}\frac{1}{O_n^l}}{1+\frac{z_n\gamma_n}{f_n}\sum_{l\in\mathscr{A}_n}\frac{1}{O_n^l}} \\
&= \frac{a_n^{0*}\frac{z_n\gamma_n}{f_n}+T_n^*\frac{z_n\gamma_n}{f_n}\sum_{l\in\mathscr{A}_n}\frac{1}{O_n^l}}{1+\frac{z_n\gamma_n}{f_n}\sum_{l\in\mathscr{A}_n}\frac{1}{O_n^l}} \\
&= \frac{T_n^*+T_n^*\frac{z_n\gamma_n}{f_n}\sum_{l\in\mathscr{A}_n}\frac{1}{O_n^l}}{1+\frac{z_n\gamma_n}{f_n}\sum_{l\in\mathscr{A}_n}\frac{1}{O_n^l}} \\
&= T_n^*,
\end{aligned}
\tag{3.37}
$$

a contradiction to the optimality of $(\mathbf{a}_n^*, T_n^*)$. Thus, for all optimal solutions, we have $\sum\limits_{m\neq n} a_m^k O_m^k + a_n^{k*} O_n^k = T_n^*$.

Proposition 3.2 can be explained as follows. Imagine HNs as containers and compare offloading tasks to HNs to pouring water to containers, then the execution time is just like the water level. To keep the highest water level as low as possible, water should be poured to containers such that the water level of every container maintains the same. It is just like the famous water-filling algorithm for power allocation problem in wireless communications.

Proposition 3.3 *For TN n, the solution* $\mathbf{a}_n$ *to problem (3.10) for a given* $\mathbf{A}_{-n}$ *is given by (3.39), which is a continuous function. Here,* $\mathscr{A}_n$ *is the set of active HNs of TN n, i.e., the HNs to which TN n offloads subtasks, and can be represented as*

$$\mathscr{A}_n = \left\{ k \in \mathscr{K} \mid \sum_{m \neq n} a_m^k O_m^k < \frac{1 + \sum_{l \in \mathscr{A}_n} \frac{\sum_{m \neq n} a_m^l O_m^l}{O_n^l}}{1 + \frac{z_n \gamma_n}{f_n} \sum_{l \in \mathscr{A}_n} \frac{1}{O_n^l}} \frac{z_n \gamma_n}{f_n} \right\}. \tag{3.38}$$

$$a_n^k(\mathbf{A}_{-n}) = \begin{cases} \dfrac{1 + \sum\limits_{l \in \mathscr{A}_n} \frac{\sum\limits_{m \neq n} a_m^l O_m^l}{O_n^l}}{1 + \frac{z_n \gamma_n}{f_n} \sum\limits_{l \in \mathscr{A}_n} \frac{1}{O_n^l}}, & k = 0 \\ 0, & \sum\limits_{m \neq n} a_m^k O_m^k \geq \dfrac{1 + \sum\limits_{l \in \mathscr{A}_n} \frac{\sum\limits_{m \neq n} a_m^l O_m^l}{O_n^l}}{1 + \frac{z_n \gamma_n}{f_n} \sum\limits_{l \in \mathscr{A}_n} \frac{1}{O_n^l}} \dfrac{z_n \gamma_n}{f_n} \\ \frac{1}{O_n^k} \left(\frac{z_n \gamma_n}{f_n} - \sum\limits_{m \neq n} a_m^k O_m^k \right) & \\ \quad - \dfrac{\frac{1}{O_n^k} \frac{z_n \gamma_n}{f_n} \sum\limits_{l \in \mathscr{A}_n} \frac{1}{O_n^l} \left(\frac{z_n \gamma_n}{f_n} - \sum\limits_{m \neq n} a_m^l O_m^l \right)}{1 + \frac{z_n \gamma_n}{f_n} \sum\limits_{l \in \mathscr{A}_n} \frac{1}{O_n^l}}, & \sum\limits_{m \neq n} a_m^k O_m^k < \dfrac{1 + \sum\limits_{l \in \mathscr{A}_n} \frac{\sum\limits_{m \neq n} a_m^l O_m^l}{O_n^l}}{1 + \frac{z_n \gamma_n}{f_n} \sum\limits_{l \in \mathscr{A}_n} \frac{1}{O_n^l}} \dfrac{z_n \gamma_n}{f_n} \end{cases}. \tag{3.39}$$

Proof Let $(\boldsymbol{a}_n^*, T_n^*)$ be the optimal solution to (3.10) and define $\mathscr{A}_n := \{k \in \mathscr{K} | a_n^{k*} > 0\}$. Then, according to Proposition 3.2, for every HN $k \in \mathscr{A}_n$, we have $\sum\limits_{m \neq n} a_m^k O_m^k + a_n^{k*} O_n^k = a_n^{0*} \frac{z_n \gamma_n}{f_n} = \big(1 - \sum\limits_{l \in \mathscr{A}_n} a_n^{l*}\big) \frac{z_n \gamma_n}{f_n}$.

Following the same derivation of Proposition 3.2, for every $k \in \mathscr{A}_n$, we have

$$\begin{aligned} a_n^{k*} =& \frac{1}{O_n^k} \left(\frac{z_n \gamma_n}{f_n} - \sum_{m \neq n} a_m^k O_m^k \right) \\ & - \frac{\frac{1}{O_n^k} \frac{z_n \gamma_n}{f_n} \sum_{l \in \mathscr{A}_n} \frac{1}{O_n^l} \left(\frac{z_n \gamma_n}{f_n} - \sum_{m \neq n} a_m^l O_m^l \right)}{1 + \frac{z_n \gamma_n}{f_n} \sum_{l \in \mathscr{A}_n} \frac{1}{O_n^l}}. \end{aligned} \tag{3.40}$$

Additionally,

$$
\begin{aligned}
& a_n^{k*} > 0 \\
\Leftrightarrow & \frac{z_n\gamma_n}{f_n} - \sum_{m\neq n} a_m^k O_m^k > \frac{\frac{z_n\gamma_n}{f_n}\sum_{l\in\mathscr{A}_n}\frac{1}{O_n^l}\left(\frac{z_n\gamma_n}{f_n} - \sum_{m\neq n} a_m^l O_m^l\right)}{1+\frac{z_n\gamma_n}{f_n}\sum_{l\in\mathscr{A}_n}\frac{1}{O_n^l}} \\
\Leftrightarrow & \frac{z_n\gamma_n}{f_n} + \left(\frac{z_n\gamma_n}{f_n}\right)^2 \sum_{l\in\mathscr{A}_n}\frac{1}{O_n^l} - \sum_{m\neq n} a_m^k O_m^k \\
& - \frac{z_n\gamma_n}{f_n}\sum_{m\neq n} a_m^k O_m^k \sum_{l\in\mathscr{A}_n}\frac{1}{O_n^l} \\
& > \left(\frac{z_n\gamma_n}{f_n}\right)^2 \sum_{l\in\mathscr{A}_n}\frac{1}{O_n^l} - \frac{z_n\gamma_n}{f_n}\sum_{l\in\mathscr{A}_n}\frac{\sum_{m\neq n} a_m^l O_m^l}{O_n^l} \\
\Leftrightarrow & \frac{z_n\gamma_n}{f_n} - \sum_{m\neq n} a_m^k O_m^k - \frac{z_n\gamma_n}{f_n}\sum_{m\neq n} a_m^k O_m^k \sum_{l\in\mathscr{A}_n}\frac{1}{O_n^l} \\
& > -\frac{z_n\gamma_n}{f_n}\sum_{l\in\mathscr{A}_n}\frac{\sum_{m\neq n} a_m^l O_m^l}{O_n^l} \\
\Leftrightarrow & \sum_{m\neq n} a_m^k O_m^k < \frac{1+\sum_{l\in\mathscr{A}_n}\frac{\sum_{m\neq n} a_m^l O_m^l}{O_n^l}}{1+\frac{z_n\gamma_n}{f_n}\sum_{l\in\mathscr{A}_n}\frac{1}{O_n^l}}\frac{z_n\gamma_n}{f_n}
\end{aligned} \tag{3.41}
$$

For $k \notin \mathscr{A}_n$,

$$
\begin{aligned}
& \sum_{m\neq n} a_m^k O_m^k \geq a_n^{0*}\frac{z_n\gamma_n}{f_n} \\
\Leftrightarrow & \sum_{m\neq n} a_m^k O_m^k \geq \left(1 - \sum_{l\in\mathscr{A}_n} a_n^{l*}\right)\frac{z_n\gamma_n}{f_n} \\
\Leftrightarrow & \sum_{m\neq n} a_m^k O_m^k \geq \left(1 - \frac{\frac{z_n\gamma_n}{f_n}\sum_{l\in\mathscr{A}_n}\frac{1}{O_n^l} - \sum_{l\in\mathscr{A}_n}\frac{\sum_{m\neq n} a_m^l O_m^l}{O_n^l}}{1+\frac{z_n\gamma_n}{f_n}\sum_{l\in\mathscr{A}_n}\frac{1}{O_n^l}}\right)\frac{z_n\gamma_n}{f_n} \\
\Leftrightarrow & \sum_{m\neq n} a_m^k O_m^k \geq \frac{1+\sum_{l\in\mathscr{A}_n}\frac{\sum_{m\neq n} a_m^l O_m^l}{O_n^l}}{1+\frac{z_n\gamma_n}{f_n}\sum_{l\in\mathscr{A}_n}\frac{1}{O_n^l}}\frac{z_n\gamma_n}{f_n}
\end{aligned} \tag{3.42}
$$

Together, this yields

$$\mathscr{A}_n = \{k \in \mathscr{K} \mid \sum_{m\neq n} a_m^k O_m^k < \frac{1 + \sum_{l\in\mathscr{A}_n} \frac{\sum_{m\neq n} a_m^l O_m^l}{O_n^l}}{1 + \frac{z_n\gamma_n}{f_n}\sum_{l\in\mathscr{A}_n}\frac{1}{O_n^l}} \frac{z_n\gamma_n}{f_n}\}. \tag{3.43}$$

The set $\mathscr{A}_n$ in Proposition 3.3 is defined in an implicit form, which is not desirable. Proposition 3.4 in the following gives an explicit formulation of $\mathscr{A}_n$ and shows that $\mathscr{A}_n$ is unique.

Proposition 3.4 *There is exactly one set $\mathscr{A}_n$, and it is of the form*

$$\mathscr{A}_n = \{k \in \mathscr{K} \mid \sum_{m\neq n} a_m^k O_m^k < \frac{1 + \sum_{l=1}^{k} \frac{\sum_{m\neq n} a_m^l O_m^l}{O_n^l}}{1 + \frac{z_n\gamma_n}{f_n}\sum_{l=1}^{k}\frac{1}{O_n^l}} \frac{z_n\gamma_n}{f_n}\}, \tag{3.44}$$

in the case of $\sum\limits_{m\neq n} a_m^1 O_m^1 \leq \sum\limits_{m\neq n} a_m^2 O_m^2 \leq \cdots \leq \sum\limits_{m\neq n} a_m^K O_m^K$.[1]

Proposition 3.4 can be explained as follows. $\sum\limits_{m\neq n} a_m^k O_m^k$ is the additional time consumption incurred by other TNs except n and can be interpreted as the current water level of container k. Proposition 3.4 shows that we can choose containers in the ascending order of the current water level until we cannot lower the water level any more. From Proposition 3.4, it is straightforward to show that the optimal solution to problem (3.10) is also unique.

Proposition 3.5 *The POST game G_2 has an NE.*

Proof Define $\mathscr{F}(\mathbf{A}) = \mathbf{a}_1(\mathbf{A}_{-1}) \otimes \mathbf{a}_2(\mathbf{A}_{-2}) \otimes \cdots \otimes \mathbf{a}_N(\mathbf{A}_{-N})$, where $\otimes$ is the Cartesian product. From Lemma 2, we know that $\mathscr{F}(\mathbf{A}) : \mathscr{S} \to \mathscr{S}$ is continuous, where $\mathscr{S} = \mathscr{S}_1 \otimes \mathscr{S}_2 \otimes \cdots \otimes \mathscr{S}_N$ is convex and compact ($\mathscr{S}_1, \mathscr{S}_2 \cdots, \mathscr{S}_N$ are polytopes). According to the *Brouwer's fixed point theorem*, $\mathscr{F}(\mathbf{A})$ has at least one fixed point on $\mathscr{S}$, which is an NE point of the POST game [36].

Among methods of finding NE points of GNEP, Gauss–Seidel-type method is the most popular in practice because its rationale is particularly simple to grasp [36]. Based on this, we design a distributed task offloading algorithm, namely POST, to obtain an NE of the POST game, shown in Algorithm 2.

Consider a slotted time structure, and every task offloading time slot is further divided into multiple decision slots. To begin with, every TN chooses local computing, i.e., $\mathbf{a}_n(0) = [1, 0, \ldots, 0]$ (line 2). Then, at every decision slot t:

[1] For TN n and HN k such that $b_n^k = 0$, we have $\sum\limits_{m\neq n} a_m^K O_m^k = \infty$.

Algorithm 2 POST algorithm

1: **initialization**:
2: every TN n chooses local computing, i.e., $\mathbf{a}_n(0) = [1, 0, \ldots, 0]$.
3: **end initialization**
4: **repeat** for every TN n and every decision slot in parallel:
5: broadcast the parameters message to the achievable HNs.
6: receive the measurements of $\sum_{m \neq n} a_m^k O_m^k$, O_n^k from the achievable HNs.
7: calculate the optimal solution to problem (3.10) according to (3.23)–(3.44).
8: **if** $T_n(t) - T_n^*(t) > \epsilon$ **then**
9:     send RTU message to HNs for contending for the strategy update opportunity.
10:     **if** not permitted to update the task offloading strategy **then**
11:         update the strategy $\mathbf{a}_n(t+1) = \mathbf{a}_n^*(\mathbf{A}_{-n}(t))$ for next slot.
12:     **else**
13:         maintain the current strategy $\mathbf{a}_n(t+1) = \mathbf{a}_n(t)$ for next slot.
14:     **end if**
15: **else**
16:     maintain the current strategy $\mathbf{a}_n(t+1) = \mathbf{a}_n(t)$ for next slot.
17: **end if**
18: **until** END message is received from the HNs.

1. *TNs broadcasting parameters to HNs*: the TNs simultaneously broadcast the parameters message to the achievable HNs, i.e., the HNs which they can access (line 5). The message informs the HNs of the task information, transmit power, and so on.
2. *HNs broadcasting measurements to TNs*: the HNs first estimate the channel conditions and calculate the measurements of current processing time, i.e., $\sum_{m \neq n} a_m^k O_m^k$, and potential extra processing time O_n^k. Then, they broadcast the measurements back to the TNs (line 6).
3. *TNs calculating the optimal solution to problem (3.10) and contending for the strategy update opportunity*: based on the measurements from HNs, the TNs can calculate the optimal solution to problem (3.10) according to (3.23)–(3.44) (line 7), and then decide whether to contend for the strategy update opportunity (line 8–9). To be specific, if $T_n(t) - T_n^*(t) > \epsilon$, i.e., TN n can further reduce its cost by ϵ, TN n sends a request to update (RTU) message to an available HN to contend for the task offloading strategy update opportunity.
4. *HNs deciding the updated TN and broadcasting the update-permission (UP) message to TNs*: once receiving the RTU messages, the HNs collaborate with each other to decide the TN which wins the update opportunity. For example, this can be realized by designating the update order in advance. After deciding the updated TN, the HNs broadcast the UP message to TNs to inform them that which TN is permitted to update its strategy.
5. *TNs updating the task offloading strategies*: If TN n is permitted to update its task offloading strategy, then it updates its strategy as $a_n(t+1) \in b_n(\mathbf{a}_{-n}(t))$ at the next decision slot (line 11). Otherwise, it maintains the current strategy at the next decision slot (line 13).

If no TNs wants to update the current strategy, i.e., no RTU messages are sent to the HNs, the HNs will broadcast the END message to all TNs and the algorithm terminates (line 18).

3.5 Performance Evaluation

3.5.1 Price of Anarchy

We first make some theoretical analysis of the proposed solutions. In game theory, price of anarchy (PoA) is usually used to evaluate the efficiency of an NE solution. It answers the question that how far is the overall performance of an NE from the socially optimal solution. PoA is defined as the ratio of the maximum social welfare, i.e., the total utility, achieved by a centralized optimal solution, to the social welfare, achieved by the worst case equilibrium [34]. Here, since we consider the cost function, we define the PoA as the ratio of the system-level average delay, achieved by the worst case equilibrium, over the system-level average delay, achieved by the centralized optimal solution. In this case, the smaller the PoA, the better the performance is.

To be specific, let Γ be the set of NEs of the POMT game G_1 and $\mathbf{a}^* = \{a_1^*, a_2^*, \ldots, a_N^*\}$ be the centralized optimal solution that minimizes the system-level average delay. Then, the PoA is defined as

$$PoA = \frac{\max_{\mathbf{a}\in\Gamma} \sum_{n\in\mathcal{N}} T_n(\mathbf{a})}{\sum_{n\in\mathcal{N}} T_n(\mathbf{a}^*)}. \tag{3.45}$$

For the POMT game G_1, we have the following proposition.

Proposition 3.6 *For the POMT game G, the PoA of the system-level average delay satisfies that*

$$1 \le PoA \le \frac{\sum_{n=1}^{N} \min\{O_{\max}, \frac{z_n\gamma_n}{f_n}\}}{\sum_{n=1}^{N} \min\{O_{n,\min}, \frac{z_n\gamma_n}{f_n}\}}, \tag{3.46}$$

where $O_{\max} =: \max_{k\in\mathcal{K}} \sum_{n\in\mathcal{N}} O_n^k$ *and* $O_{n,\min} =: \min_{k\in\mathcal{K}} O_n^k$.

Proof Let $\bar{\mathbf{a}} \in \Gamma$ be an arbitrary NE of the game G_1. Since the centralized optimal solution $\mathbf{a}^*$ minimizes the system-level average delay, we have that $PoA \ge 1$.

For the NE solution $\bar{\mathbf{a}}$, if $\bar{\mathbf{a}}_n > 0$, then we have $T_n(\bar{\mathbf{a}}) = \sum_{m\neq n} O_m^{a_m} I_{\{a_m=a_n\}} + O_n^{a_n} \le \max_{k\in\mathcal{K}} \sum_{n\in\mathcal{N}} O_n^k = O_{\max}$. $O_{\max}$ is the maximum delay that TN n can achieve in the worst case where all TNs choose to offload tasks to the same HN. Moreover, if $\frac{z_n\gamma_n}{f_n} < O_{\max}$ and $\bar{a}_n > 0$, then the system-level average delay can be further reduced by letting TN n switch to the local computing. This is because that switching to the local computing will not increase the execution time of other TNs. As a result, we know that

$$T_n(\bar{\mathbf{a}}) \leq \min\left\{O_{\max}, \frac{z_n\gamma_n}{f_n}\right\}. \tag{3.47}$$

For the centralized optimal solution $\mathbf{a}^*$, if $a_n^* > 0$, then we have $T_n(\mathbf{a}^*) \geq O_n^{a_n^*} \geq \min_{k\in\mathcal{K}} O_n^k = O_{n,\min}$. $O_{n,\min}$ is the minimum delay that the TN n can achieve in the best case where only TN n chooses HN computing while the other TNs choose local computing, i.e., no competition. Moreover, if $\frac{z_n\gamma_n}{f_n} < O_{n,\min}$ and $a_n^* > 0$, then the system-level average delay can be further reduced by letting TN n switch to the local computing. As a result, we know that

$$T_n(\mathbf{a}^*) \geq \min\left\{O_{n,\min}, \frac{z_n\gamma_n}{f_n}\right\}. \tag{3.48}$$

In conclusion,

$$1 \leq PoA = \frac{\max_{\mathbf{a}\in\Gamma}\sum_{n\in\mathcal{N}} T_n(\mathbf{a})}{\sum_{n\in\mathcal{N}} T_n(\mathbf{a}^*)} \leq \frac{\sum_{n=1}^N \min\{O_{\max}, \frac{z_n\gamma_n}{f_n}\}}{\sum_{n=1}^N \min\{O_{n,\min}, \frac{z_n\gamma_n}{f_n}\}}. \tag{3.49}$$

Similarly, for the POST game G_2, we have the following proposition.

Proposition 3.7 *For the POST game G_2, the PoA in terms of the system average delay satisfies that*

$$1 \leq PoA \leq \frac{\sum_{n=1}^N \min\{O_{n,\max}^{\text{sum}}, \frac{z_n\gamma_n}{f_n}\}}{\sum_{n=1}^N \min\{O_{n,\min}, \frac{z_n\gamma_n}{f_n}\}} \times (1 + |\mathcal{K}_n|), \tag{3.50}$$

where $O_{n,\max}^{\text{sum}} \triangleq \max_{k\in\mathcal{K}} \sum_{n\in\mathcal{N}} O_n^k$, $O_{n,\min} \triangleq \min_{k\in\mathcal{K}} O_n^k$, $\mathcal{K}_n = \{k \in \mathcal{K} | b_n^k = 1\}$ and $|\mathcal{K}_n|$ is the cardinality of set $\mathcal{K}_n$.

Proof Let $\bar{\mathbf{A}} \in \Gamma$ be an arbitrary NE of the POST game G_2. Since the centralized optimal solution $\mathbf{A}^*$ minimizes the system-wide average delay, we have that $PoA \geq 1$.

For the NE solution $\bar{\mathbf{A}}$, if TN n chooses to offload part of its task to HNs, i.e., $\bar{\mathbf{a}}_n \neq \mathbf{e}_1$, where $\mathbf{e}_1$ is a unit $1 \times (K+1)$ vector with the first entry being 1, then we have $T_n(\bar{\mathbf{a}}_n, \bar{\mathbf{A}}_{-n}) = \sum_{m\neq n} \bar{a}_m^k O_m^k + \bar{a}_n^k O_n^k \leq \max_{k\in\{k\in\mathcal{K}|b_n^k=1\}} \sum_{n\in\mathcal{N}} O_n^k = O_{n,\max}^{\text{sum}}$. $O_{n,\max}^{\text{sum}}$ is the maximum delay that TN n can achieve in the worst case where all TNs choose to offload entire tasks to the same HN. Moreover, if $\frac{z_n\gamma_n}{f_n} < O_{n,\max}^{\text{sum}}$ and $\bar{\mathbf{a}}_n \neq \mathbf{e}_1$, then the system-wide average delay can be further reduced by letting TN n switch to the local computing, i.e., $\bar{\mathbf{a}}_n = \mathbf{e}_1$. This is because that switching to the local computing will not increase the execution time of other tasks. As a result, we know that

$$T_n(\bar{\mathbf{a}}_n, \bar{\mathbf{A}}_{-n}) \leq \min\left\{O_{n,\max}^{\text{sum}}, \frac{z_n\gamma_n}{f_n}\right\}. \tag{3.51}$$

For the centralized optimal solution $\mathbf{A}^*$, if $\mathbf{a}_n^* \neq \mathbf{e}_1$, then we have $T_n(\mathbf{a}_n^*, \mathbf{A}_{-n}^*) \geq \min_{\mathbf{a}_n \in \mathcal{S}_n} T_n(\mathbf{a}_n, \mathbf{0}) \geq \frac{1}{1+|\mathcal{K}_n|}\min\{O_{n,\min}, \frac{z_n\gamma_n}{f_n}\}$, where $O_{n,\min} \triangleq \min_{k\in\mathcal{K}} O_n^k$, $\mathcal{K}_n = \{k \in \mathcal{K}|b_n^k = 1\}$ and $|\mathcal{K}_n|$ is the cardinality of set $\mathcal{K}_n$. It indicates the minimum delay that the TN n can achieve in the best case where only TN n chooses HN computing while the other TNs choose local computing, i.e., no competition. In this case, by offloading tasks to at most $|\mathcal{K}_n|$ HNs, the delay can be reduced to at most $\frac{1}{1+|\mathcal{K}_n|}\min\{O_{n,\min}, \frac{z_n\gamma_n}{f_n}\}$.

In conclusion,

$$\begin{aligned} 1 \leq PoA &= \frac{\max_{\mathbf{a}\in\Gamma} \sum_{n\in\mathcal{N}} T_n(\mathbf{a})}{\sum_{n\in\mathcal{N}} T_n(\mathbf{a}^*)} \\ &\leq \frac{\sum_{n=1}^{N} \min\{O_{n,\max}^{\text{sum}}, \frac{z_n\gamma_n}{f_n}\}}{\sum_{n=1}^{N} \min\{O_{n,\min}, \frac{z_n\gamma_n}{f_n}\}} \times (1 + |\mathcal{K}_n|). \end{aligned} \tag{3.52}$$

3.5.2 System Average Delay

We then evaluate the performance of the proposed POMT algorithm and POST algorithm via simulations. Consider an 80 m × 80 m square area, and divide it into 4 × 4, totally 16 grids. Deploy a HN in the center of every grid, which not only possesses rich communication resource, but also provides strong computation capability. Every HN has a coverage range of 20 m, a bandwidth of 5 MHz, and a computation capability of 5 GHz [15]. The TNs are randomly scattered in the area. To account for the heterogeneity of TNs and tasks: (1) the size and processing density of tasks are randomly chosen from [500, 5000] KB and [500, 3000] cycle/bit, which are in accordance with the real measurements in practice[2] [43]; (2) the computation capability of TNs is randomly selected from the set [0.8, 0.9, 1.0, 1.1, 1.2] GHz[3] [13]. Besides, the transmission power of all TNs is set as 100 mW and the channel noise power is set to be −100 dBm. The channel gain g_n^k is modeled by $(d_n^k)^{-\alpha}$, where d_n^k is the distance between TN n and HN k

[2]For example, the data size and processing density of face recognition application are about 5000 KB, 3000 cycle/bit respectively [43].

[3]Here, we model the TNs as various mobile devices with limited computation capability. For example, smartphones with ARM Cortex-A8 processors are with about 1 GHz clock speed [18].

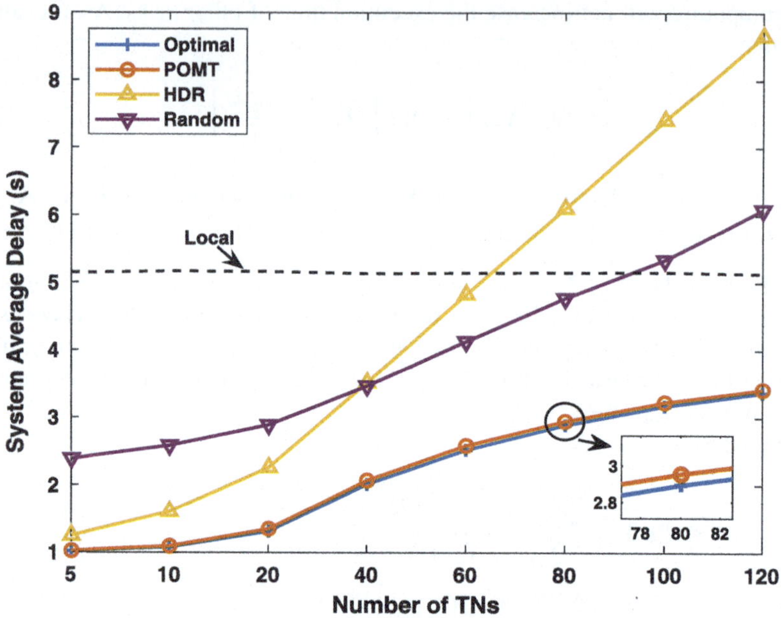

Fig. 3.2 System average delay with different number of TNs (non-splittable tasks)

and $\alpha = 4$ is the pathloss factor [15]. Unless stated otherwise, ϵ is chosen as 10^{-3}. All simulation results are averaged over 500 simulation rounds, if not specified.

Figure 3.2 compares the proposed POMT algorithm with the following baseline solutions in terms of the system average delay:

- local computing (Local): every TN chooses to process its task on local device.
- HDR offloading (HDR): every TN chooses to offload its task to the HN with the highest data rate (HDR). It imitates the myopic behavior of individuals.
- random offloading (Random): every TN randomly chooses to process its task on local device or offload it to a randomly selected HN.
- optimal offloading (Optimal): the centralized optimal solution in terms of system-level average delay is obtained [44], utilizing the Cross Entropy method. The Cross Entropy method is an efficient method for finding near-optimal solutions to complex combinatorial optimization problems [45].

As illustrated in Fig. 3.2, the system average delay increases as the number of TNs increases, except for the local computing. This is because that the communication resources and computation capabilities of HNs become insufficient when the number of TNs increases. As a result, less TNs choose to offload tasks to HNs rather than to process tasks on local device, and this results in the rise of the system average delay. While for the local computing, it does not rely on the resources of HNs, and thus it keeps unchanged. Besides, it can be observed that our POMT algorithm can always achieve the near-optimal system average delay and thus reduce

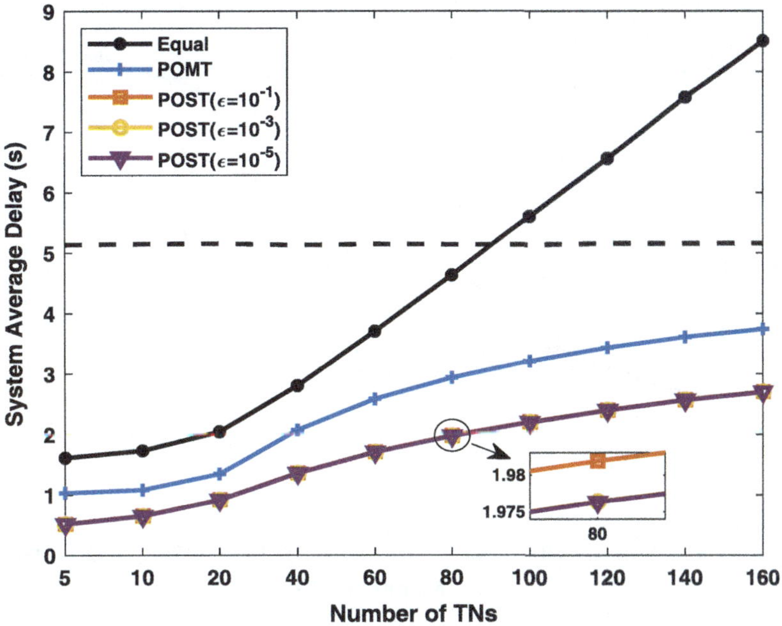

Fig. 3.3 System average delay with different number of TNs, under different schemes and ϵ (splittable tasks)

27–74%, 40–67%, and 49–53% system average delay than the local computing, HDR offloading, and random offloading, respectively. This confirms the derivation above, i.e., the PoA of system average delay is bounded.

Figure 3.3 compares the POST algorithm with the following baselines in terms of the system average delay:

- local computing (Local): every TN processes its task on local device.
- equal scheduling (Equal): every TN divides its task into multiple subtasks with equal size, and all subtasks are processed on local TN and all available HNs in parallel. This is a heuristic solution which utilizes all available computation resources.
- POMT: this solution can be taken as a worst case or upper bound for our problem.

As shown in Fig. 3.3, the system average delay increases with the number of TNs increasing. This is because that the communication resources and computation resources become insufficient when the number of TNs rises. Therefore, less TNs will offload tasks to HNs, and this results in the rise of system average delay. Besides, it can be observed that the POST algorithm can always offer the best performance and reduce over 50% and 25% delay than the Equal scheme and the POMT scheme, under the ideal condition, i.e., no extra time cost for dividing tasks and assembling results. It is also interesting to note that the value of ϵ almost has no impact on the system average delay. Specially, as ϵ is smaller than the magnitude

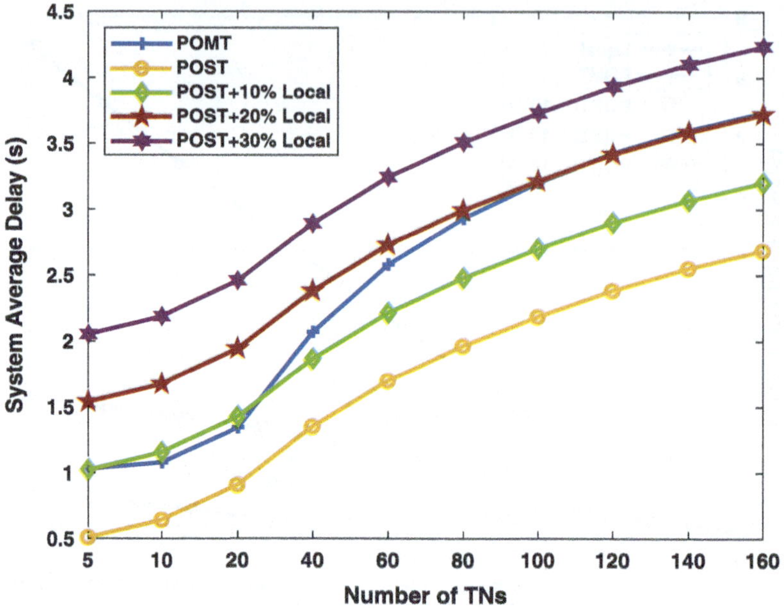

Fig. 3.4 System average delay with different number of TNs, under different schemes and extra time costs (splittable tasks)

of 10^{-3}, the system average delay achieved by the POST algorithm maintains the same. This indicates that the POST algorithm possesses a good robustness.

To further study the impact of extra time cost for dividing tasks and assembling results, Fig. 3.4 plots the POST algorithm with different extra time costs. Typically, the time cost for dividing tasks and assembling results is proportional to the complexity of tasks, and inversely proportional to the computation capability of devices. Thus, the time cost for dividing tasks and assembling results is set to be proportional to that of local computing, such as 10% (POST+10% Local), 20% (POST+20% Local), and 30% (POST+30% Local).

It is interesting to notice that the parallel processing (POST) does not always perform better than the non-parallel processing (POMT). It depends on the extra time cost for dividing tasks and assembling results and the network size. Generally speaking, the smaller the extra time cost for dividing tasks and assembling results is, the more advantageous the POST algorithm is. While, the larger the number of TNs is, the more advantageous the POST algorithm is. For example, if the time cost of dividing tasks and assembling results is 10% of that of local computing, the POST algorithm can obtain lower system average delay than the POMT scheme, as long as the number of TNs is larger than 20. And, the performance gain becomes larger as the number of TNs rises. While, if dividing tasks and assembling results take up 20% of the time cost of local computing, the POST algorithm performs worse than the POMT scheme in terms of the system average delay, when the number of TNs is less than 100. While, the performance of POST algorithm and the performance of POMT scheme almost maintain the same, as TNs are more than 100.

It is also worth noting that it seems that whatever the number of TNs, the POMT scheme almost cannot achieve better performance than the POST+10% Local scheme. It may be claimed that dividing tasks into subtasks and offloading them to multiple HNs for parallel processing is a better choice, as long as the extra time cost for dividing tasks and assembling results takes up less than 10% of the time cost of local computing.

3.5.3 Number of Beneficial TNs

The beneficial TNs are those who can reduce its delay compared with local computing, via offloading tasks to HNs, i.e., $\{n \in \mathcal{N} | T_n < T_n^0\}$. This metric answers the question that how many TNs can benefit from computation offloading, i.e., reduce its delay via computation offloading. Therefore, it can not only reflect the individual-level delay performance but also depict the satisfaction level of TNs to the pairing result from individual respects.

As shown in Fig. 3.5, the number of beneficial TNs achieved by the POMT algorithm increases with the number of TNs increasing and levels off. Moreover,

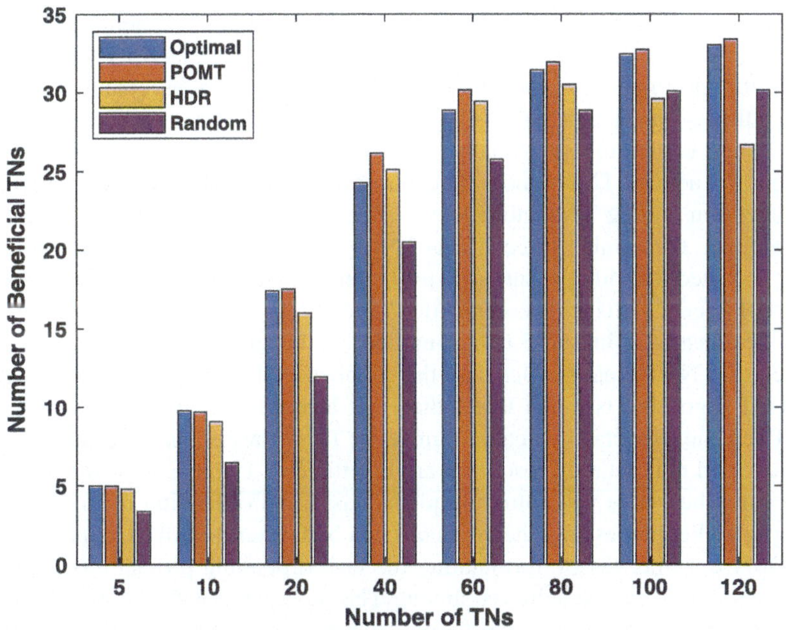

Fig. 3.5 Number of beneficial TNs with different number of TNs (non-splittable tasks)

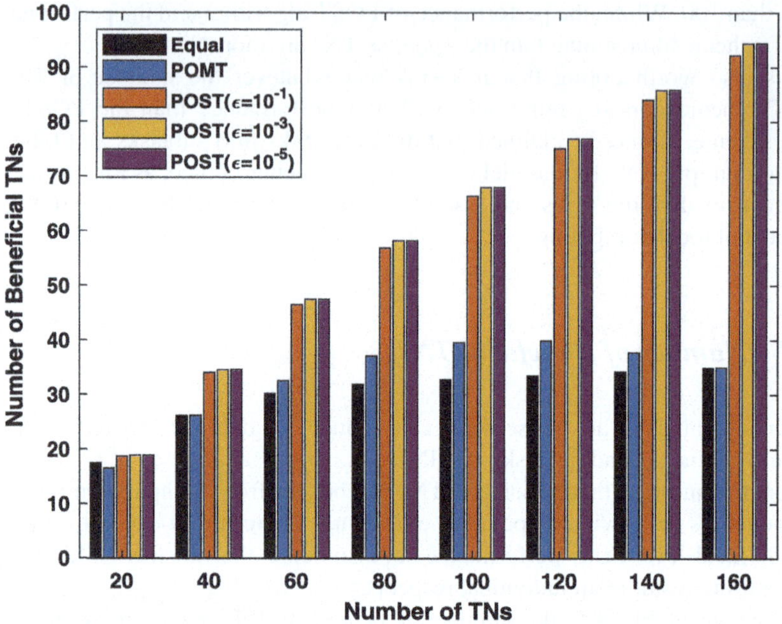

Fig. 3.6 Number of beneficial TNs with different number of TNs (splittable tasks)

the POMT algorithm can always achieve more number of beneficial TNs than other three schemes, especially when the number of TNs is large. This is because that, in the case of large number of TNs, there exists serious contention among TNs for the resources of HNs. While all TNs offloading tasks to HNs will further worsen such contention and finally damage each other's interests, and thus the number of beneficial TNs reduces. In contrast, the POMT algorithm can coordinate the contention among TNs and achieve more beneficial TNs until the network is saturated, i.e., the number of beneficial TNs levels off. As for the optimal solution, it may sacrifice individuals' interests for better system performance, and thus less TNs would get benefits from computation offloading.

As demonstrated in Fig. 3.6, the number of beneficial TNs increases as the number of TNs increases. Besides, the number of beneficial TNs achieved by the POST algorithm is much larger than that achieved by the Equal scheme and POMT scheme, especially when the number of TNs is large, i.e., the communication resources and computation resources are insufficient. The reason is two-folds: (1) by dividing the whole task into multiple subtasks and offloading them to multiple HNs for parallel processing, the efficiency can be further improved, compared with the POMT scheme; (2) by formulating the task offloading problem as a game, a better negotiation can be achieved among TNs, compared with the Equal scheme. Therefore, more TNs can benefit from computation offloading and the POST algorithm can also offer much better performance in terms of the system average delay, compared with the Equal scheme and POMT scheme. These advantages are much more evident when there are a large number of TNs in the network, because there will be more serious contention among TNs. Again, as can be observed, ϵ almost has no influence on the algorithm performance in terms of the number of beneficial TNs, as long as ϵ is small enough.

3.5.4 Convergence

Figure 3.7 demonstrates the number of decision slots required for the POMT algorithm and the POST algorithm to converge, with different number of TNs and under different ϵ. As shown, the number of decision slots increases sub-linearly with the number of TNs, regardless of ϵ. This indicates that both the POMT algorithm and the POST algorithm scale well with the size of network. Besides, the number of decision slots increases slightly as ϵ decreases. It is because that it takes more efforts for the POST algorithm to get closer to the NE (when ϵ is smaller). But, as ϵ is small enough, i.e., below the magnitude of 10^{-3}, the number of decision slots does not increase any more. Further, the number of decision slots required for the POST algorithm is usually more than the number of decision slots required for the POMT algorithm. The reason behind is intuitive and straightforward. As the competition among TNs gets more serious and complex for dividing original tasks into multiple subtasks, it is more difficult to reach an equilibrium among TNs.

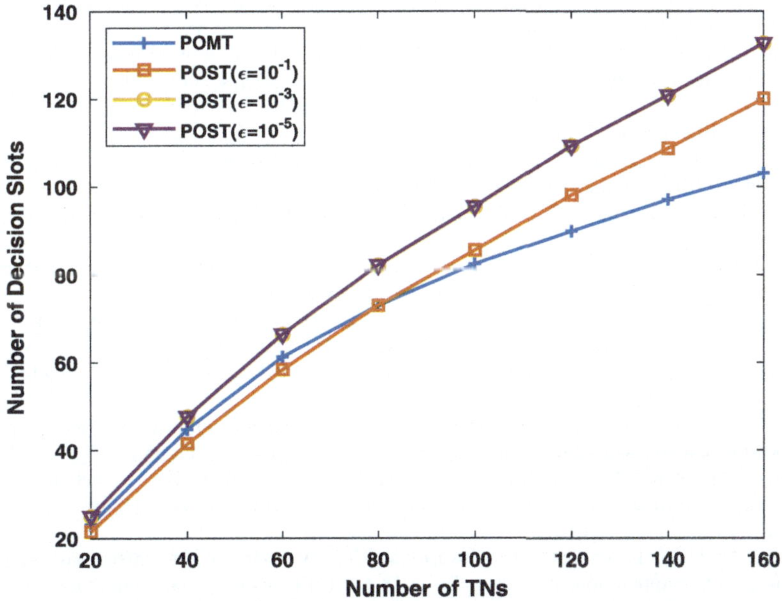

Fig. 3.7 Number of decision slots with different number of TNs

3.6 Conclusion

In this chapter, we investigated the task offloading problem in MTMH fog networks under the case of non-splittable tasks and splittable tasks, respectively. We proposed a game theory based analytical framework and a corresponding distributed algorithm for each. To be specific, we first introduced the general MTMH fog networks and the corresponding task offloading problem for non-splittable tasks and splittable tasks, respectively. Then, we discussed the fundamentals of game theory and how to apply it to formulate and tackle the task offloading problems introduced above. More specifically, the POMT game based on potential games and the POST game based on GNEP were formulated and studied thoroughly. The corresponding distributed algorithms, namely POMT and POST, were also designed. Finally, extensive simulations were conducted to demonstrate the performance of the proposed POMT and POST algorithms.

References

1. Bonomi F, Milito R, Zhu J, Addepalli S (2012) Fog computing and its role in the internet of things. In: Proceedings of the first edition of the MCC workshop on Mobile cloud computing. ACM, New York, pp 13–16
2. Chiang M, Zhang T (2016) Fog and IoT: an overview of research opportunities. IEEE Internet Things J 3(6):854–864
3. Openfog reference architecture technical paper. https://www.openfogconsortium.org/wp-content/uploads/OpenFog_Reference_Architecture_2_09_17-FINAL-1.pdf
4. Chen S, Xu H, Liu D, Hu B, Wang H (2014) A vision of IoT: applications, challenges, and opportunities with China perspective. IEEE Internet Things J 1(4):349–359
5. Chen S, Hu J, Shi Y, Zhao L (2016) LTE-V: a TD-LTE-based V2X solution for future vehicular network. IEEE Internet Things J 3(6):997–1005
6. Lu N, Cheng N, Zhang N, Shen X, Mark JW (2014) Connected vehicles: solutions and challenges. IEEE Internet Things J 1(4):289–299
7. Chen N, Yang Y, Zhang T, Zhou M-T, Luo X, Zao JK (2018) Fog as a service technology. IEEE Commun Mag 56(11):95–101
8. Ali M, Riaz N, Ashraf MI, Qaisar S, Naeem M (2018) Joint cloudlet selection and latency minimization in fog networks. IEEE Trans Ind Inf 14(9):4055–4063
9. Yang Y, Liu Z, Yang X, Wang K, Hong X, Ge X (2019) POMT: paired offloading of multiple tasks in heterogeneous fog networks. IEEE Internet Things J Early Access
10. Meng X, Wang W, Zhang Z (2017) Delay-constrained hybrid computation offloading with cloud and fog computing. IEEE Access 5:21355–21367
11. Dinh TQ, Tang J, La QD, Quek TQS (2017) Offloading in mobile edge computing: task allocation and computational frequency scaling. IEEE Trans Commun 65(8):3571–3584
12. Yang Y, Wang K, Zhang G, Chen X, Luo X, Zhou M-T (2018) MEETS: maximal energy efficient task scheduling in homogeneous fog networks. IEEE Internet Things J 5(5):4076–4087
13. Nowak D, Mahn T, Al-Shatri H, Schwartz A, Klein A (2018) A generalized nash game for mobile edge computation offloading. In: 2018 6th IEEE international conference on mobile cloud computing, services, and engineering (MobileCloud). IEEE, Piscataway, pp 95–102

14. Zhao P, Tian H, Qin C, Nie G (2017) Energy-saving offloading by jointly allocating radio and computational resources for mobile edge computing. IEEE Access 5:11255–11268
15. Chen X, Jiao L, Li W, Fu X (2016) Efficient multi-user computation offloading for mobile-edge cloud computing. IEEE/ACM Trans Netw 24(5):2795–2808
16. Jošilo S, Dán G (2017) A game theoretic analysis of selfish mobile computation offloading. In: IEEE INFOCOM 2017-IEEE conference on computer communications. IEEE, Piscataway, pp 1–9
17. Jošilo S, Dán G (2018) Selfish decentralized computation offloading for mobile cloud computing in dense wireless networks. IEEE Trans Mob Comput 18(1):207–220
18. Shah-Mansouri H, Wong VWS (2018) Hierarchical fog-cloud computing for IoT systems: a computation offloading game. IEEE Internet Things J 5(4):3246–3257
19. Yang L, Zhang H, Li X, Ji H, Leung VCM (2018) A distributed computation offloading strategy in small-cell networks integrated with mobile edge computing. IEEE/ACM Trans Netw 26(6):2762–2773
20. Tran TX, Pompili D (2019) Joint task offloading and resource allocation for multi-server mobile-edge computing networks. IEEE Trans Veh Technol 68:856–868
21. Pham Q., Leanh T., Tran N.H., Park B.J., Hong C.S. (2018) Decentralized computation offloading and resource allocation for mobile-edge computing: a matching game approach. IEEE Access 6:75868–75885
22. Li X, Liu Y, Ji H, Zhang H, Leung VCM (2019) Optimizing resources allocation for fog computing-based internet of things networks. IEEE Access 7:64907–64922
23. Zhao S, Yang Y, Shao Z, Yang X, Qian H, Wang C-X (2018) FEMOS: fog-enabled multitier operations scheduling in dynamic wireless networks. IEEE Internet Things J 5(2):1169–1183
24. Yang Y, Zhao S, Zhang W, Chen Y, Luo X, Wang J (2018) DEBTS: delay energy balanced task scheduling in homogeneous fog networks. IEEE Internet Things J 5(3):2094–2106
25. de Souza VBC, Ramírez W, Masip-Bruin X, Marín-Tordera E, Ren G, Tashakor G (2016) Handling service allocation in combined fog-cloud scenarios. In: 2016 IEEE international conference on communications (ICC). IEEE, Piscataway, pp 1–5
26. Pang A-C, Chung W-H, Chiu T-C, Zhang J (2017) Latency-driven cooperative task computing in multi-user fog-radio access networks. In: 2017 IEEE 37th international conference on distributed computing systems (ICDCS). IEEE, Piscataway, pp 615–624
27. Deng R, Lu R, Lai C, Luan TH, Liang H (2016) Optimal workload allocation in fog-cloud computing toward balanced delay and power consumption. IEEE Internet Things J 3(6):1171–1181
28. Oueis J, Strinati EC, Barbarossa S (2015) The fog balancing: load distribution for small cell cloud computing. In: 2015 IEEE 81st vehicular technology conference (VTC Spring). IEEE, Piscataway, pp 1–6
29. Oueis J, Strinati EC, Sardellitti S, Barbarossa S (2015) Small cell clustering for efficient distributed fog computing: a multi-user case. In: 2015 IEEE 82nd vehicular technology conference (VTC Fall). IEEE, Piscataway, pp 1–5
30. Muñoz O, Pascual-Iserte A, Vidal J (2015) Optimization of radio and computational resources for energy efficiency in latency-constrained application offloading. IEEE Trans Veh Technol 64(10):4738–4755
31. Li N, Martinez-Ortega J-F, Rubio G (2018) Distributed joint offloading decision and resource allocation for multi-user mobile edge computing: a game theory approach. Preprint. arXiv:1805.02182
32. Liu Z, Yang X, Yang Y, Wang K, Mao G. (2018) DATS: dispersive stable task scheduling in heterogeneous fog networks. IEEE Internet Things J 6(2): 3423–3436
33. Xing H, Liu L, Xu J, Nallanathan A (2018) Joint task assignment and wireless resource allocation for cooperative mobile-edge computing. In: 2018 IEEE international conference on communications (ICC). IEEE, Piscataway, pp 1–6
34. Han Z, Niyato D, Saad W, Başar T, Hjørungnes A (2012) Game theory in wireless and communication networks: theory, models, and applications. Cambridge University Press, Cambridge

35. Lã QD, Chew YH, Soong B-H (2016) Potential game theory: applications in radio resource allocation. Springer, Berlin
36. Facchinei F, Kanzow C (2010) Generalized Nash equilibrium problems. Ann Oper Res 175(1):177–211
37. Han Z, Gu Y, Saad W (2017) Matching theory for wireless networks. Springer, Berlin
38. Zhang N, Zhang S, Zheng J, Fang X, Mark JW, Shen X (2017) QoE driven decentralized spectrum sharing in 5G networks: potential game approach. IEEE Trans Veh Technol 66(9):7797–7808
39. Zheng J, Cai Y, Liu Y, Xu Y, Duan B, Shen X (2014) Optimal power allocation and user scheduling in multicell networks: base station cooperation using a game-theoretic approach. IEEE Trans Wirel Commun 13(12):6928–6942
40. Monderer D, Shapley LS (1996) Potential games. Games Econom Behav 14(1):124–143
41. Liu Z, Yang Y, Wang K, Shao Z, Zhang J (2019) POST: parallel offloading of splittable tasks in heterogeneous fog networks. IEEE Internet Things J (under review)
42. Hoheisel T (2009) Mathematical programs with vanishing constraints. J Optim Theory Appl 142(3):501–532
43. Kwak J, Kim Y, Lee J, Chong S (2015) DREAM: dynamic resource and task allocation for energy minimization in mobile cloud systems. IEEE J Sel Areas Commun 33(12):2510–2523
44. Liu Z, Yang Y, Zhou M-T, Li Z (2018) A unified cross-entropy based task scheduling algorithm for heterogeneous fog networks. In: Proceedings of the 1st ACM international workshop on smart cities and Fog computing. ACM, New York, pp 1–6
45. Rubinstein RY, Kroese DP (2004) The cross entropy method: a unified approach to combinatorial optimization, Monte-Carlo simulation (Information science and statistics). Springer, New York

Chapter 4
Fog-Enabled Multi-Robot System

4.1 Introduction

Robots are used widely in many fields now. They bring us lots of convenience in daily lives, save huge manpower in factories, and help to complete many mission-possible tasks in some cases. However, in robot applications, it may suffer issues of high cost, large power consumption, and low efficiency. For example, in a logistic center employing huge number of automated guided vehicles (AGVs), high performance algorithms must be used to avoid collisions. If all computing tasks generated by a AGV are executed by itself, then it requires each AGV to have high-performance computing unit and large capacity battery. While, this will introduce issues of high cost and low efficiency.

An effective solution to robot application issues is to employ fog computing, a new computing paradigm that is emerging in recent years. By the definition of OpenFog [1, 2], fog computing utilizes all possible resources (vertically) along the cloud-to-thing continuum and (horizontally) across networks and vertical industries to perform computing, storage, control, and communication, and services can be deployed anywhere in the network. In the paradigm of fog computing, routers, switches, file servers, gateways, and base stations may serve as fog nodes (FNs) for computing, storage, and control [3, 4]. With fog computing, if a robot is connected to a fog network, then part of the heavy computing tasks can be offloaded to one or a group of FNs that have much higher computing capacity, so to save onboard unit cost, computing time, and energy consumption. Moreover, if multiple robots are connected each other through a fog network, then the robots may collaborate each other and improve general intelligence if machine learning algorithms are employed and training data are shared. Cloud may offload computing tasks of robot applications, too [4]. However, the robot applications may suffer intermittent cloud connectivity. Cloud offloading also introduces much longer time delay and consumes much more communication resources.

Y. Yang et al., *Fog-Enabled Intelligent IoT Systems*,
https://doi.org/10.1007/978-3-030-23185-9_4

This chapter serves to illustrate how fog computing helps robot applications in saving cost and improve efficiency. Three cases are introduced: robot simultaneous location and mapping (SLAM), robot smart factory applications, and robot fleet formation. The applications may involve one single robot or multiple robots. Meanwhile, it shows that how fog computing resource can be sold and service can be provided via a virtual market in a blockchain platform. With economic stimulus, one might be more willing to share fog computing power to meet others needs including robot applications.

4.2 Simultaneous Location and Mapping

In the real world, robots are often needed to explore a prior-unknown environment. For example, when people are buried in a collapsed building in an earthquake, and the space/condition does not allow rescue person/animal to enter, an advanced method is to send a suitable size and shape robot to detect the location of the lives, environment, and the living conditions like oxygen level and temperature, etc. Mapping information is critical to rescue planning in this case; thus, it requires an explorer robot to construct a map of the space. Meanwhile, to do so it needs to know its location and orientation in the map, too. In robotics, SLAM is a robot to perform construction of an environment map and estimate its location concurrently.

4.2.1 System Architecture

SLAM is a process consisting of concurrent construction of a map of the environment and estimation of the position and orientation of the robot. SLAM is very important in terms of, first, to support many tasks like path planning and intuitive human operator visualization, and second, to limit the error committed in estimating the state of the robot. At the IEEE Robotics and Automation Conference in 1986, researchers proposed to apply the estimation-theoretic methods in mapping and localization. Initially researchers used SLAM in robotics field to build a robot environment map and estimate the location of the robot based on the map without any prior knowledge.

Assuming a robot with sensors (e.g., camera) moves in an unknown environment, for purpose of convenience, it notes the time as discrete moments $t = \{t_1, t_2, \dots, t_k\}$, and the location as $x = \{x_1, x_2, \dots, x_k\}$. The location x forms the trajectory of the robot. Supposing the map consists of a number of signposts, at each time moment, the sensor will measure some signposts and have the observed data. Noting the map signposts as $y = \{y_1, y_2, \dots, y_N\}$, SLAM is to estimate x and y, using the movement measurement and sensing data. Figure 4.1 illustrates the SLAM principles.

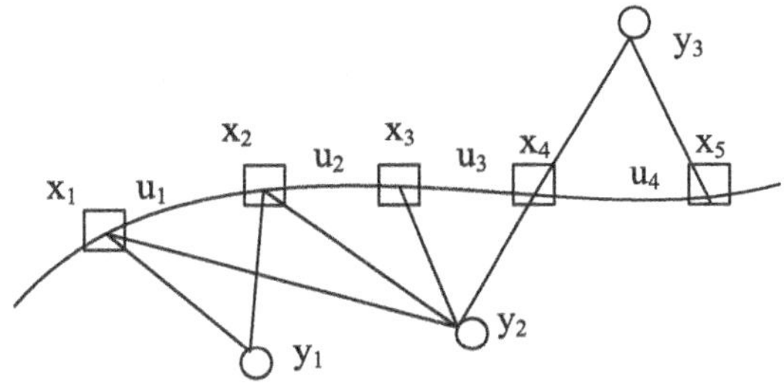

Fig. 4.1 Principles of robot SLAM

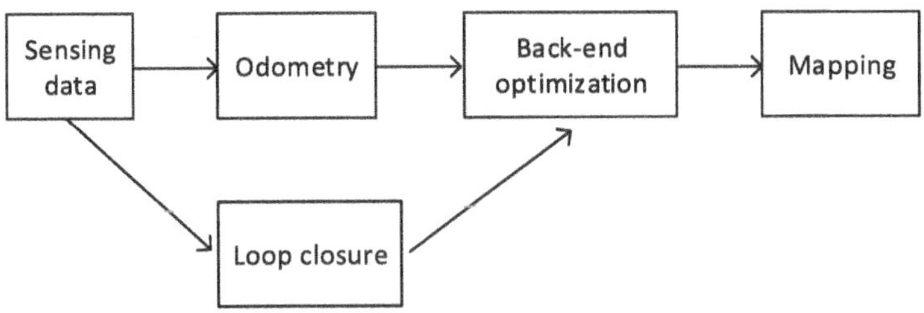

Fig. 4.2 Robot SLAM steps

Robot SLAM often uses two types of sensors, i.e., cameras and LIDAR. It names the SLAM using only camera sensor as visual SLAM (vSLAM). Cameras have rich vision information and low-cost hardware; so many robot SLAM adopts vSLAM. High resolution LIDAR is usually expensive; however, some applications also use relatively cheap LIDAR. As shown in Fig. 4.2, a classic robot SLAM involves four steps, i.e., odometry, back-end optimization, loop closure, and mapping [1]. Robot SLAM involves a number of steps, i.e., collecting data by sensors, front-end odometry, back-end optimization, loop closure, and mapping [1]. Sensors are used to collect environment information. Odometry is to estimate the movement of the robot based on the sensing data and optimization is to reduce noise and distortions in the first two steps to improve accuracy. Loop closure is to recognize the intersecting points of the movement, then to estimate the real environment topology. In these steps, optimization has the heaviest computation tasks.

4.2.2 Technical Challenges

Generally, robot SLAM needs to be low-cost, low-power consumption, accurate, and speedy. However, these requirements restrict each other mutually. First, tens or hundreds of rescue robots may be used in one saving campaign so robot cost is a big concern. Popular SLAM sensors include laser radars and cameras; however, although generally laser radars perform better in accuracy, they are much expensive than usual cameras, so the latter is more suitable for massive rescue robots. Second, accurate mapping and localization are very important to rescue path planning, to achieve this it needs high-performance computing unit, especially at the step of optimization that involves many advanced algorithms, and this of course, is contradictory to the low-cost requirement aforementioned. Third, rescue robots use battery; hence, battery life is a key consideration. To reduce computing tasks of robots, e.g., using low-performance algorithms is an effective way to save robot energy consumption; however, it may lead to inaccurate SLAM. Fourth, as all know, time is critical in many rescue cases, as it requires robots to move as quick as possible and perform fast SLAM, imposing much pressure to the onboard computing unit. Similarly, low-complexity algorithms may be used to save SLAM time; however, a sequence is that the SLAM accuracy may be affected. Fifth, in a large area where multiple robots are used, it requires collaborations between the robots in SLAM and to merge the maps finally, thus a network is needed and one robot needs to be a leader in SLAM and to merge maps, this leader robot, of course, will consume additional energy. However, network is usually not available in disaster scenarios.

4.2.3 *Fog-Enabled Solution*

Fog computing is effective to alleviate robot SLAM issues. Figure 4.3 illustrates multiple robots are connected to a fog network, by which each robot's SLAM can be offloaded to the FN. As the robots are mobile, they connect to the fog network via wireless access, possibly Wi-Fi or 4G/5G. The fog network may connect to a cloud but it is not a must, particularly in disaster scenarios where cloud is not available. Computation offloading can be done by one FN or multiple FNs jointly. Note that the fog network may consist of homogenous or heterogeneous networks, either wire or wireless networks.

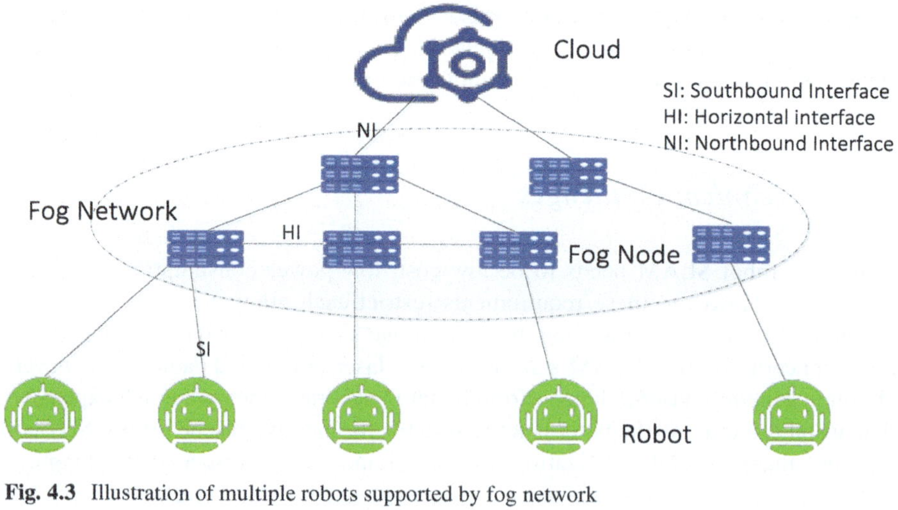

Fig. 4.3 Illustration of multiple robots supported by fog network

The function interface between robots and FNs is called southbound interface which defines information of the data plane and control plane. The data plane carries key frames and poses information which are processed by robots or FNs. The control plane carries commons which manage the whole work flows such as initializing, stopping, getting key frames, and setting poses. FNs can communicate with each other via horizontal interface. Multiple FNs can collaborate and realize CoMP (coordinated multiple points), distribute storage system, or distribute computing system to get more powerful capabilities of communication, storage, and computing. The horizontal interface carries the control and data information supporting coordinating functions. By utilizing FNs or fog network which are much closed to robots, the latency of robots processing can be shortened to a 10ms level. FNs can communicate with cloud via northbound interface. In robot SLAM, except collecting sensing data, other computation steps can be offloaded to the fog network. In case cameras are used, the video streams require high data rate wireless transmission, e.g., several to tens Mbps. On the other hand, the front-end odometry needs (relatively) much lower computing capacity than other steps, and the output data rate of the front-end processing may be as low as tens kbps. Therefore, it is proposed to offload only steps after odometry to the fog network.

Figure 4.4 illustrates the scheme for fog-enabled robot SLAM. It assumes that the robots use cameras due to cost consideration and supports collaborations of multiple robots SLAM. First, a FN initializes odometry process on a robot and then the robot starts front-end processing like to abstract key frames and positing itself. The robot reports key frame streams, which are the main visual capture representing the environment, to the fog network to optimize the map. The fog network sends back each robot the position information and the robot will correct position drift accumulated. When cloud is available, it can merge maps from different robots. If connection to cloud is not possible, then one master FN may merge maps from different robots.

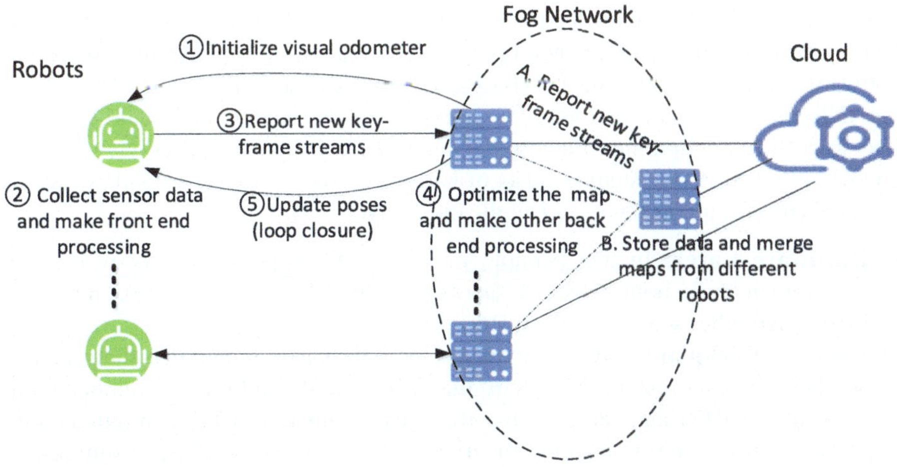

Fig. 4.4 Fog-enabled robot SLAM

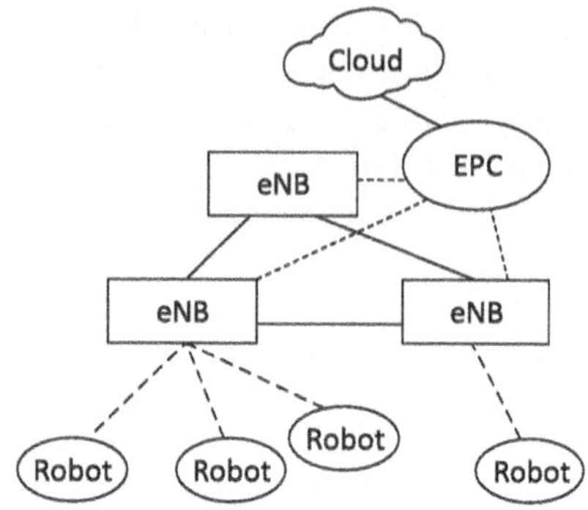

Fig. 4.5 Example OpenLTE fog network supporting robot SLAM

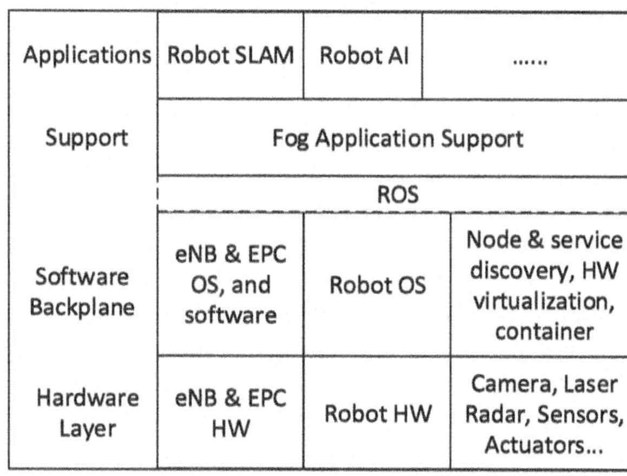

Fig. 4.6 Architecture view of fog-enabled robot SLAM with support of OpenLTE

To illustrate the fog-enabled robot SLAM more intuitively, the system uses OpenLTE as communication means connecting robots and fog network [6]. Figure 4.5 shows example fog network supporting robot SLAM and employing OpenLTE. Figure 4.6 shows its architecture view. OpenLTE is a software-defined LTE system that runs on the general purpose processor (GPP) platform. It includes software-defined eNodeB (SDN-eNB) and software-defined EPC (SDN-EPC). Both the eNB and EPC could be a miniPC, a notebook, or a server, and could serve as FNs to perform computing offloading of robot SLAM. In architecture view, a fog-enabled robot SLAM system consists of hardware layer, software backplane layer, fog application support layer, and applications layer.

- **Hardware Layer**: In the example of OpenLTE supported fog network, the hardware include both eNB/EPC hardware, the robot hardware, and cameras, laser radar, other sensors and actuators, etc.
- **Software Backplane Layer**: It manages the FNs hardware and FNs themselves, so they include operation system like Linux and Android, communication software, and FN and services discovery, etc. With OpenLTE, communication software include eNB and EPC functions. FN discovery is a basic requirement

of fog network, by which neighbor FNs are discovered and then it is possible to perform resource sharing and distributed computing [7]. This layer may include hardware virtualization and container. With virtualization technology, computing and storage resource on a FN can be sliced for different applications with good isolations. Another important component of backbone software layer is robot operation system (ROS), which is a robotics middleware providing services for heterogeneous computing clusters such as hardware abstraction, low-level device control, implementation of commonly used functionality, message-passing between processes, and package management. It has a modular design, and each module can be deployed on different FNs in distributed manner. A function module may publish message that others may subscribe and then data flows among different nodes. ROS enables data exchange between robots and FNs easily.

- **Fog Application Support Layer**: This layer supports fog applications and has a broad spectrum. It supports security, storage, message, runtime engine, etc.
- **Application Layer**: A bound of applications that may be supported by fog networks include robot SLAM, AI, etc.

Figure 4.7 shows the function view of fog-enabled robot SLAM. The function modules include task process, communication, storage, deployment, security, and interfaces. Considering low latency requirement and stream-oriented processing, it may apply real-time streaming processing framework such as Storm and Spark. Taking Storm as an example, the task process can be composed of application, task scheduler, cluster coordination, supervisor, and worker. To be compatible with the legacy mechanism, deployment manager may adopt ROS as well. Storage manager can support distributed storage system and centralized databases. Distributed storage systems are suitable for these delay insensitive applications while demanding large storage. Centralized databases are suitable for delay sensitive systems. One can select different storage scheme through deployment manager. Security manager can manage different security functions such as authentication, encryption, privacy protection, and so on. The southbound interface provides information exchange of control plane, data plane, and manage plane between FNs and robots. The horizontal interface supplies information exchange between FNs to realize fog net-based collaborative functions. The northbound interface is to exchange information between FNs and cloud to support delay-insensitive, computing-intensive and huge-storage tasks.

Table 4.1 shows how the challenges and requirements of fog-enabled robot SLAM can be met.

A testbed of fog-enabled robot SLAM is shown in Fig. 4.8. We employed rescue robot with relatively low price (shown in Fig. 4.8a). The robot was equipped with a mobile phone and a fish eye camera was attached on the mobile phone to capture environment video streaming. When there was no fog network connection, SLAM computing was done by the mobile phone on the robots. However, the computing capacity of the mobile phone is relatively weak.

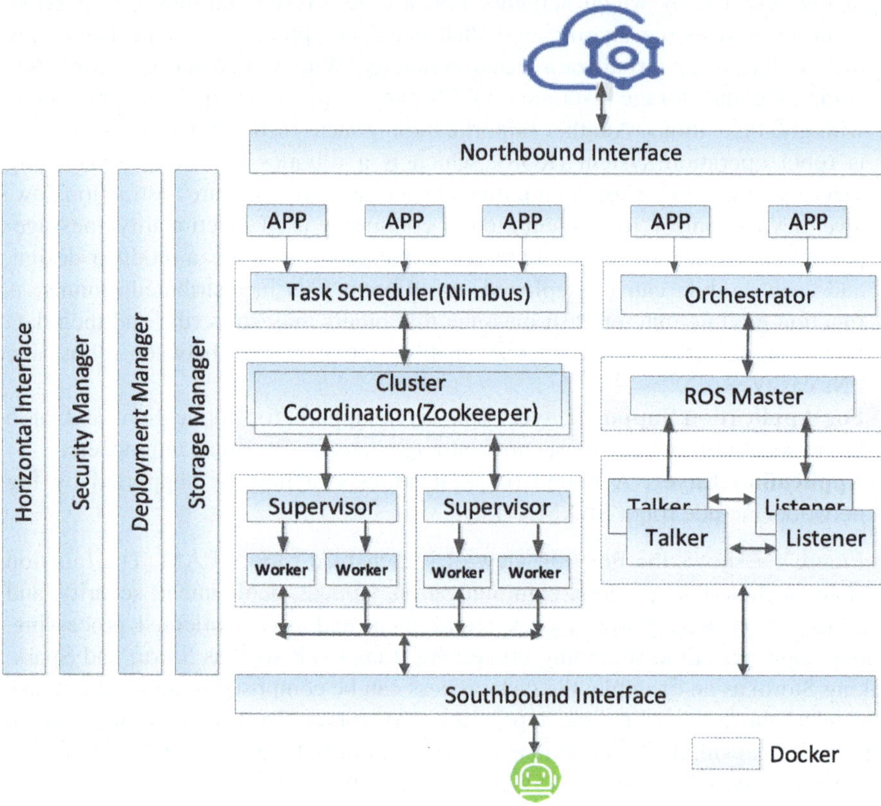

Fig. 4.7 Function view of fog-enabled robot SLAM

The FN was a server installed software-defined OpenLTE components including eNB and EPC. USRP B210 was used to convert LTE RF signal to baseband. The FN processed baseband signal, implemented eNB and EPC functions. LTE downlink and uplink bandwidth were 10 and 5 MHz, and speed were 35 and 15 Mbps, respectively. As the FN was computer in nature, it was easy to carry out information processing like SLAM computation offloading. We installed customized SIM card on the robot mobile phone so it could communicate with the FN by LTE signal. Comparing to Wi-Fi, LTE has advantages of supporting to handover, and better QoS, etc. A smooth handover is required when a robot moves to another cell.

Two FNs with OpenLTE protocols were developed and two robots were employed in the testbed, and they formed a local LTE network without cloud connection. Both the robots and FNs installed ROS. The robots executed first two SLAM step modules, and the FNs run other SLAM step modules. One FN served as the ROS master, coordinating ROS modules on different nodes. At the fog network side, the two FNs monitored its resource real-time utilizations and exchanged this information each other. The orchestrator on the FN running ROS

Table 4.1 Proposed functions meeting challenges and requirements for fog-enabled robot SLAM

Challenges and requirements	Functions
Security	1. Robots ID authentication should be supported on SI
	2. Data information encryption should be supported on SI, NI, and HI
	3. The sensitive information included in key frames and map data should be encrypted or erased before stored into databases
Scalability	1. Task scheduler can orchestrate computation, storage, communication resources based on the task QoS requirements, status of robots, FNs, and cloud-end
	2. For delay sensitive tasks, task scheduler can assign them to robots or neighboring FNs. For delay insensitive tasks, task scheduler can assign them to cloud
	3. Task scheduler should consider computing-communication tradeoff when making decision on task scheduling
	4. The FNs can form a collaborative network to support more powerful communication, computing, and storage systems
Open	1. The information model, message primitives, and protocol stack of application layer on SI, NI, and HI should be standardized based on the common robot applications
	2. The policies of task scheduling, resources orchestration, and modules deployment can be open and defined by users
	3. All kinds of resources should be visualized by GUI
Autonomy	1. The network topology should be flexible to support different application scenarios, such as stand-alone robots, network of robot and FNs, network of robot, FNs, and cloud-end network
	2. Both robots and FNs can finish task independently
RAS	1. Robots can independently work without the connection with FNs and cloud-end
	2. Fog nodes can finish work without the connection with cloud-end
	3. Fog nodes can organize a collaborative network to support much more reliable communication, powerful communication, and huge storage
Agility	1. Machine learning functions can be deployed on FNs or cloud-end to provide wiser schedule scheme
	2. Robots can learn more experience and knowledge from big history data in FNs or cloud-end
Hierarchy	1. The solution can be one (robots only), two (robots and fog nodes), or three (robots, FNs, and cloud-end) layers
	2. Each level can work independently
	3. Fog nodes can provide distributed computing, storage solution for higher QoS applications. Each level can work independently
Programmability	1. The work flow of applications can be programmed to achieve different QoS requirements
	2. The policies of task scheduling, resources allocation, information process mechanism can be programmed by users

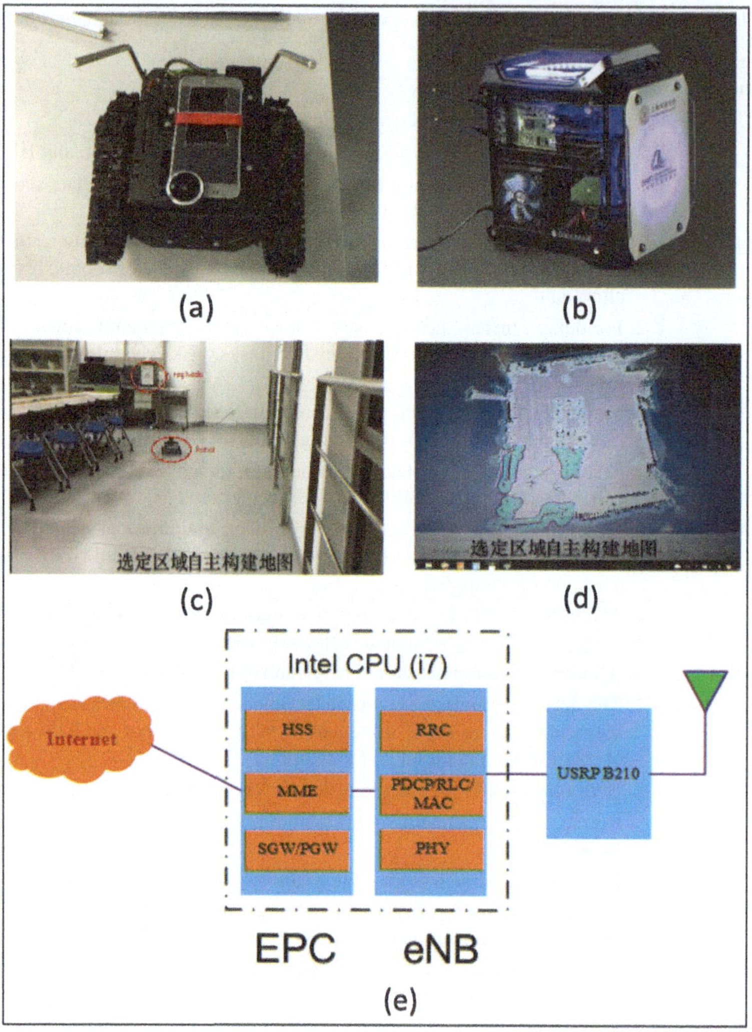

Fig. 4.8 Developed testbed of fog-enabled robot SLAM with OpenLTE. (**a**) Rescue robot with mobile phone; (**b**) FN running software-defined OpenLTE; (**c**) experiment area; (**d**) SLAM map built on FN; (**e**) illustration of the OpenLTE architecture implemented

master coordinated resource usage, if one FN was short of computing resource, then the SLAM computing tasks can be shifted to the other FN with the help of ROS.

In fog architecture view, the testbed consists of the following items:

- Hardware Layer: robot with mobile phone and fish eye camera, FN hardware include i7 CPU, USRP, RF HW, and FPGA for acceleration, hard disk for data storage, antenna, etc.

- Software Backplane Layer: ROS, Android on mobile phone, Linux Ubuntu on FN, orchestrator, resource monitor, task scheduler, etc.
- Fog Application Support: map storage, map viewing
- Application Layer: robot SLAM

Figure 4.8d illustrates SLAM mapping with fog network. It had been achieved good results in terms of speed. The process was more fluent comparing to Wi-Fi connection without fog-enabled SLAM.

4.3 Multi-Robot Smart Factory

Following the first three revolutions of mechanization, mass production, and digitization, the 4th industrial revolution, termed as Industry 4.0, has brought autonomous technologies into the industrial realm. Consequently, traditional factories are transforming into smart factories of the future with the help of recent advancement of information technologies including Internet of Things (IoT), big data, and cloud computing together with artificial intelligence and multi-robot systems. Generally, industrial robots are deployed in factories to do tedious, repetitive, or dangerous tasks, such as assembly, painting, packaging, and welding. These preprogrammed robots have been very successful in industrial applications due to their high endurance, speed, and precision in structured factory environments. To extend the functional range of these robots or to deploy them in unstructured environments, robotic technologies are further integrated with wired or wireless communication networks to form networked robots [8], which address the concerns associated with single robot by sharing the perceived data among robots and solving a task in a cooperative and/or coordinated manner [9].

Due to hardware resource constraints, industrial robots still face enormous computational and storage issues in compute intensive tasks, e.g., map reconstruction, navigation and real-time response, etc. Although multi-robot cooperative systems facilitate global cooperation through coordinated information exchange, their information processing is often limited by the number of robots in the network topology. The cooperative learning process also became complicated for the robots in the dynamic working environment like smart factory, because it is usually required for the machines and processes to communicate and negotiate with each other so as to reconfigure themselves for a flexible production of multiple types of products. Furthermore, the load balance of the network bandwidth is another challenge for robot communication in smart factory [10].

To overcome these limitations, the concept of cloud computing has been extended to multi-robot systems [11]. In such systems, cloud offers virtualized instances, platform, and software services so that both local and remote resources can be utilized to process the tasks [12]. The fourth industrial revolution introduces an ideal opportunity for the inclusion of cloud-enabled robots in a factory environment to improve productivity and reduce human intervention. Thanks to virtualization, decentralization, and real-time capability, smart factory is envisioned to be a key area for infusion of these robotic technologies, especially in automating applications such as sensing, actuating, and monitoring empowered by cloud computing and wireless sensors. Indeed, industrial cloud robotics amalgamates the design principle of robotic resources with the powerful computing and network resources, which has extended its operational capabilities and produced a shift in the modes of robotic applications from carrying out repetitive tasks towards solving more complex multiobjective problems in uncertain environments [13, 14].

In essence, smart factory is an intelligent production system that utilizes the encapsulation of manufacturing and services. It integrates communication process, computing process, and control process to meet the industrial requirements. The defining characteristics of the smart factory are visibility, connectivity, and autonomy [15]. Among them, the most fundamental one is its connectedness, for which sensors are critical to linking devices, machines, and systems to provide data needed in order to make real-time decisions. From the smallest node to the most complex robotic system, a smart factory relies on connectivity [16]. In this case, the remote cloud platform alone can hardly perceive environmental status timely and fulfill the real-time demands within the smart factory [17]. Therefore, diverse applications are required to migrate to the network edge with the help of fog computing to provide edge intelligent services [18].

In this section, we discuss different components of fog-enabled multi-robot system in smart factory and it is followed by two case studies including warehouse management and emergency management. Then, we brief the challenges to incorporate fog technology within smart factory.

4.3.1 System Architecture

With the adoption of industrial IoT and robotics technologies in the factory, the opportunity for innovation arises where the networked robots and sensors are integrated with cloud infrastructure for intelligent perception and on-demand shared resources, resulting in increased efficiency for environmental monitoring, improved supply chains, reduced waste as well as enhanced safety and speed of manufacturing. However, the rapid increases of terminal devices, massive data analytics, and processing of things gradually raise up the burdens on manufacturing cloud platform [19], making it insufficient to realize real-time requirements of smart factory. On the basis of emerging fog computing, the industrial fog applications will make the full use of computing capabilities of an intelligent equipment to achieve the global perception and autonomous distributed information processing [17]. Given that, we will first present an integrated architecture of fog-aided multi-robot system in smart factory (Fig. 4.9) as follows:

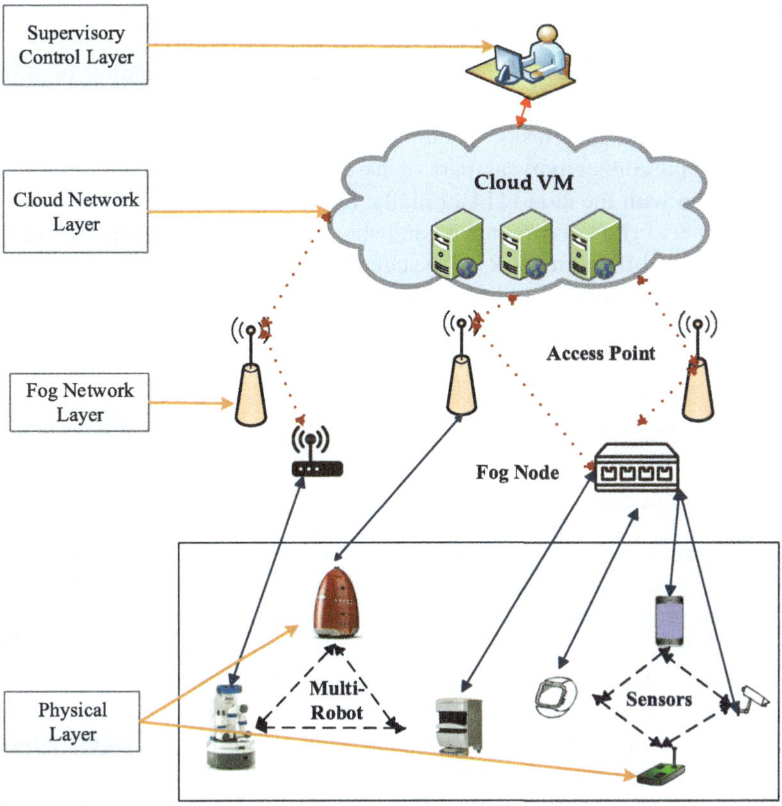

Fig. 4.9 Fog-enabled multi-robot system environment in smart factory

Physical Layer

The industrial physical layer consists of two components: wireless sensor network and robotic agents. The wireless sensor network (WSN) contains low-cost and limited-energy smart sensor devices embedded in the machines, which is able to not only communicate with each other, but also collect and analyze raw data necessary for application specific tasks. The sensors provide directive information to the robot through cloud orchestration. Based on that, several analytical computation is done and a task is assigned to robots for visiting locations and conducting jobs that are allocated to them. Here machine learning in sensors helps them to identify key situations such as suggesting modifications or requirements for actions that can be performed by robots.

The robots are the critical element that can perform actuation and help complete the tasks. Depending on the type of application, the robot is able to share its workload with the fog nodes or cloud by offloading the tasks for additional support. Robots may also get guidance provided by the supervisory control center through the cloud. Based on all of that, tasks are shared among robotic, edge, and cloud resources. As robots possess the distinctive attribute of mobility, they can plan their routes in accordance and hence choose the suitable communication platform to access the virtual resources while moving. As a result, it gains the best available bandwidth for offloading tasks to either fog nodes or cloud. Robot offloading with motion and connectivity as part of its decision-making could thus improve communication with the cloud [14]. Finally, robots are also able to form their own local ad hoc network to communicate and share information among themselves as well as assist with offloading tasks to the edge or remote cloud. Therefore, the multi-robot system is equipped to utilize both its surrounding local resources (available robots and fog nodes) and global virtual resources (virtual machines in cloud) to share workload.

Fog Network Layer

The industrial network layer consists of access points (APs), routers, or switches that enable the robotic and sensor network to communicate with each other. It also bridges the gap between cloud services and the physical layer components for data collection and processing. In this context, the network resource is defined as a smart device with Internet connection via which the robots access the cloud infrastructure. The components in network layer also have the processing and storage capabilities, playing the role as fog nodes, which help execute the latency sensitive tasks and provide real-time interaction of the system. As there are multiple APs, the robots could obtain different stream rates for communication depending on its location and AP association in accordance with IEEE 802.11 WLAN protocol, for instance. It means the bandwidth received by a robot may vary depending on the location it offloads tasks from or the robot that is offloading or the AP it is selecting. Hence, the tangible fog network layer enables the intangible information to flow freely by integrating the physical components and information entities.

Cloud Network Layer

Cloud infrastructure refers to hardware and software components such as servers, storage, and virtualization techniques that are needed to support the computing requirements of the application. It includes an abstraction layer that virtualizes resources and logically presents them to users through application program interfaces (API) and API-enabled command line or graphical interfaces. The organization of cloud typically consists of virtual machines (VMs) with shared power. They often provide functionality needed to execute the entire high-density operating systems and require massive computational capacity to handle unpredictable and complex user (robot) demands. Some notable cloud service providers are Mendix, Google App Engine, Amazon Web Services, etc. In particular, cloud services in our context are the VMs, which virtualize computing resources as the back-end components and perform on-demand computational support (for offloaded task), data storage (data collected from sensors), analytics (decision-making, verification) to aid local robots.

Supervisory Control Layer

The supervisory control layer allows networked robots to be guided/monitored by humans remotely through the cloud infrastructure. Here the information collected from sensors and action reports performed by robots are passed on to cloud via fog nodes and then made available for users to monitor through control terminals. As a result, physical layer could communicate with users/engineers in remote locations when required. In addition, possible big data analytics could also provide various statistical results to the users for the purpose of supervisory control and the users then verify/adjust system configuration accordingly.

4.3.2 *Warehouse Management*

The impact of Industry 4.0 applications is obviously being reflected in the warehouse applications, e.g., material handling including conveyers, sorters, goods to picker solution, where more operations are now moving towards running with automation support. According to recent reports, around 15% of current warehouses are mechanized and 5% are automated, but most of them still employ people in key functions. It suggests that there is a large room for the implementation of automated components like robotic agents. In fact, the inclusion of robots in the warehouse management could be traced back to early 1990s, where the approach initially started with teleoperation and later upgraded to automation [20]. Having evolved with time, it has now reached the age of cloud networked robotics, where cloud computing in robotic applications has made significant mark in the industrial domain by enhancing operations of the robots via on-demand computation and storage support.

In the context of our application, a pool of wireless sensors are deployed in static warehouse machineries (e.g., goods packing, labeling, etc.) for data collection and environmental monitoring in order to develop a common operating picture (COP). These sensors are complemented by several dynamic robots that move on-demand to perform object pickup, delivery, and drop-off. The integration of these cyber-physical systems and wireless sensors enables proper communication over networks for data sharing and automated processing of operations that start from production line all the way to delivery. Since the design of industrial operations involves numerous varieties of decision-making, the inclusion of cloud computing makes an integrated architecture of networked robots and sensors a reality, which is able to reconcile conventional warehouse problems and perform applications in a semi-automated manner with minimal human supervisory oversight, increased efficiency, and more safety and speed.

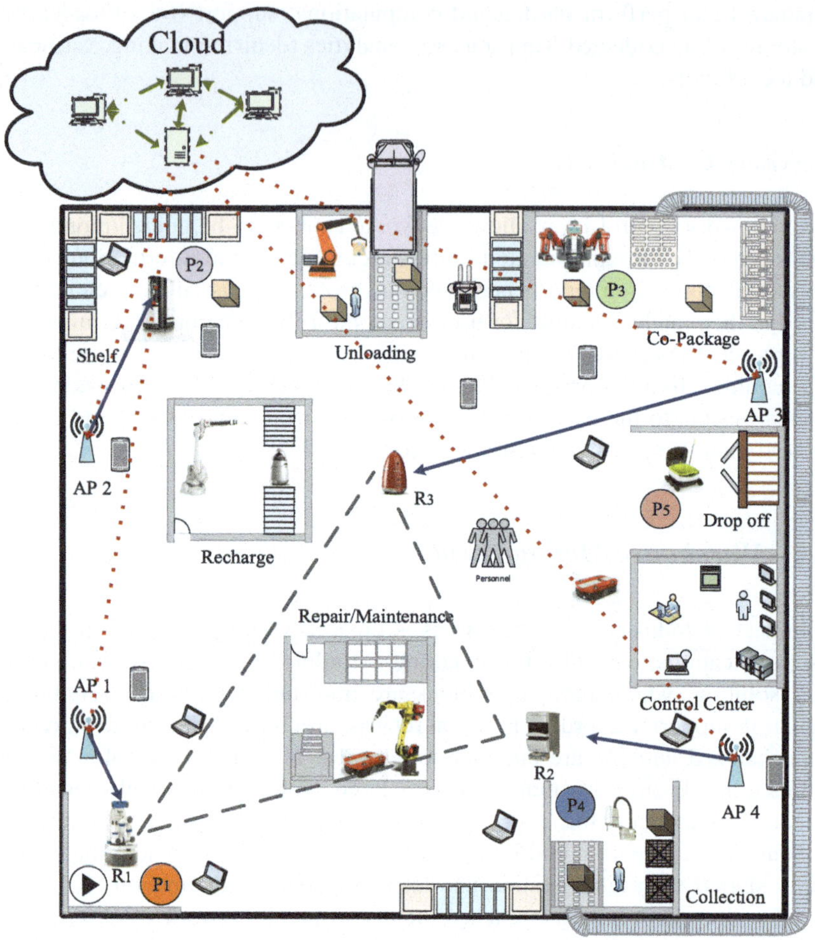

Fig. 4.10 Floor schematic of a smart warehouse environment

Specifically, we present a parcel sorting and distribution application in an automated warehouse environment as shown in Fig. 4.10. Being automated in nature, the application deals with five major steps that require the primary mobile robot to complete set of tasks necessary to carry out a parcel for delivery. In this multi-robot system, the principal/primary robot is incorporated to complete the major actuation-based tasks through interactions with different types of agents, each with a specific job such as unloading objects from trucks, co-packing, picking orders, checking inventory or shipping goods to assist. As for the supporting robots, they provide analytical and computation support while completing their own set of application related activities. Through robot-to-robot communication, the principal robot can transfer tasks locally to other available ones and complete them on-board.

Alternatively, it can also get help from the supporting robots regarding offloading a task to cloud for utilizing its ubiquitous resources. In this manner, the other available robots in this shared architecture work as a hub to provide assistance for local or cloud based offloading of tasks (if required), whereas the principal robot carries out fundamental aspects of the warehouse management application as well as necessary decision-making. The workflow of automated parcel sorting and distribution in smart warehouse is depicted in detail as Fig. 4.11.

- Parcel Request Generation Phase (stage 1): The warehouse distribution and sorting centers are equipped with sensors; therefore, the complete application will be coordinated through advanced warehouse management systems. Each machinery is capable of tracking inventory movements and progressing orders with a high degree of accuracy. As part of it, whenever a new order is set to be sorted and delivered, information regarding its location and target will be sent to the principal robot ($R1$), which in this case is a fetch and freight robot. Its primary robot, called fetch, can extend its torso to reach pickup points, while a small secondary robot, called freight, helpfully holds the tote that fetch will pick items into. Each fetch robot can have several of these smaller freight robots supporting the picking process. Moreover, due to their size they can smoothly move around and collect objects throughout the warehouse and hence been chosen as our principal multi-functioning robot. As a parcel sorting request for a new order is generated, robot gets the location and plans its path from current location ($P1$) to go to the given area ($P2$) for collecting the parcel.
- Pickup and Co-packaging Phase (stage 2): As the robot is reaching the location ($P2$) of the shelves, a mobile piece picking robot is placed in that area which uses 3D cameras for identifying objects and implements a well-defined grasping technique for collecting objects from the shelves. After the object is picked and kept in a convenient location, the update is provided to the principal robot through WSN, so that it could detect object and pick it up. The next portion of the application involves co-packaging and customization for parcel according to individual needs of the customer. In comparison to more traditional/manual procedures, the robot in this setting carries the parcel to the co-packaging center ($P3$), where the well-known robot Baxter can complete the necessary steps of packaging. During these applications, a lot of information and analytical tasks are happening in parallel. That is why local or cloud based offloading may be required for more efficient performance of the system.
- Package Collection Phase (stage 3): After the parcel is customized and co-packaged, it is ready for delivery. At that point, they are placed on conveyer belt to be sent to the collection center. As updated information is provided to the principal robot ($R1$), it moves to the collection point ($P4$) to pick up the prepared parcel. While moving, the robot needs to plan its path and communicate with the collection center to provide update to the main center. This creates the opportunities to pass on heavy computational tasks to nearby supporting robots for local computation or for assistance with offloading to the cloud.

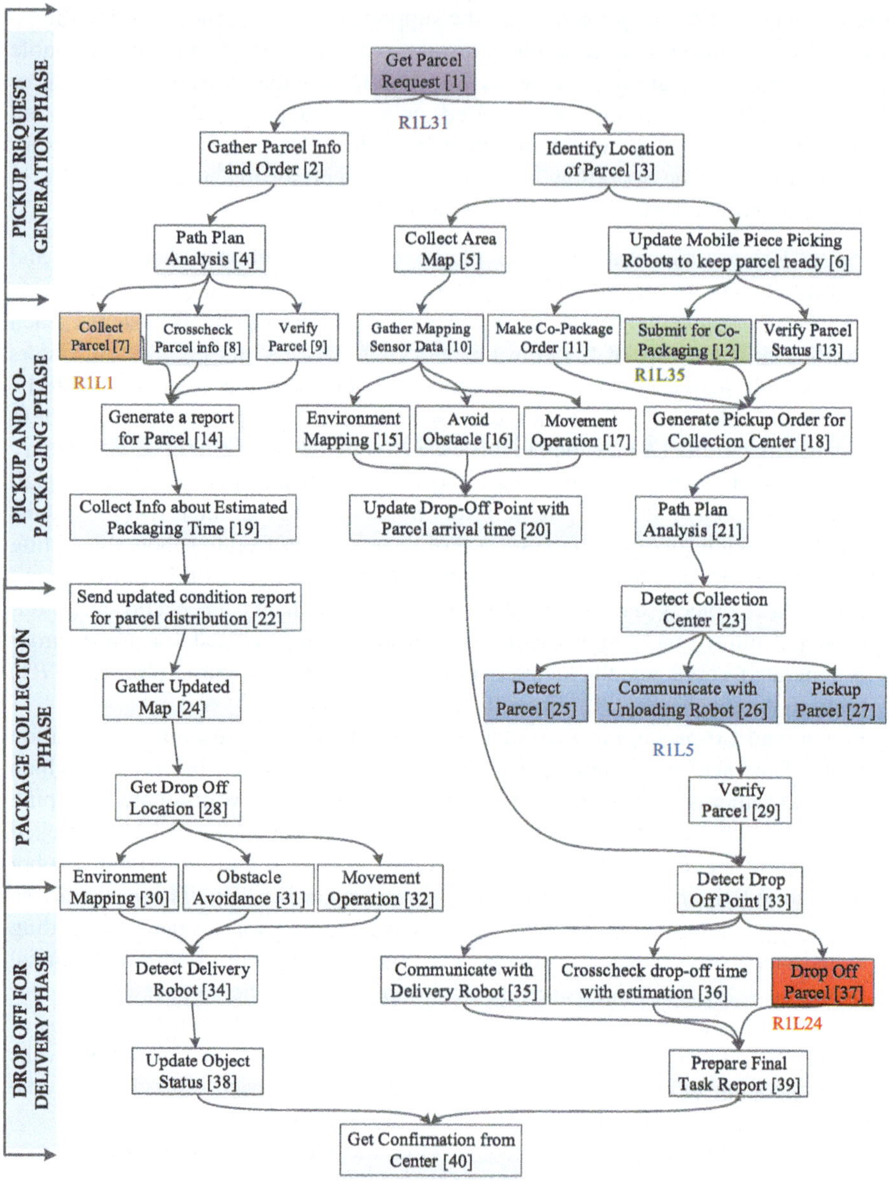

Fig. 4.11 Workflow of automated parcel sorting and distribution application in smart warehouse

- Drop-off for Delivery Phase (stage 4): As the robot reaches the collection point,' it detects the prepared parcel, and uses its own technology to pick it up. Then it updates the main center and additionally creates an order for the delivery robots from Starship technologies to be prepared for incoming parcel. At last, the robot delivers the objects in the drop-off point ($P5$) to be collected by delivery robot.

Due to the nature of the application, it is time constrained, which is why additional support from cloud and other available robots may improve the performance. Through joint collaboration of cloud and multi-robot resources, parcel distribution and sorting process in a smart warehouse is run autonomously to make parcel ready for delivery, starting from distribution to the eventual drop-off. Since task offloading is dependent on communication with the cloud, it requires Internet availability. Depending on the location of robot and the selection of AP, a robot gains access to the available bandwidth, which in turn may impact its offloading process. In this way, it is quite clear that task offloading, path planning, and AP selection are interdependent that allows robots to plan their path and select the communication link while accommodating the offloading decisions. In addition, local sharing among multiple robots has the potential of further minimizing system energy by distributing the principal robot's workload to others for either computational support or assistance with offloading to the cloud. Therefore, the collective consideration of decision-making for each task (i.e., task offloading, robot selection for offloading, task location, and AP selection) could better utilize the local resources in the multi-robot systems, and also the offloading process could be further fastened, leading to successful task completion with less energy consumption but more system performance enhancement [21]. On this basis, we formulate a four-layer joint optimization problem with the objective of minimizing the overall system energy by identifying the following key decisions: (1) which task to allocate to robot and which task to offload to the cloud, (2) which robot would offload the task to the cloud, (3) which location to complete a task, and (4) which AP to select for offloading a particular task. The simulation results suggest that the implementation of our genetic algorithm (GA) based scheme achieves near-optimal solutions and in turn improves the overall system performance for cloud networked multi-robot applications in smart factory.

Mathematically, the objective is to minimize the overall energy consumption (E_{total}) in order to meet the time constraint (T_{Deadline}) and individual energy constraint ($E_{\text{limit}}(R_r)$) of each robot (E_{R_r}), which is represented as follows:

$$\begin{aligned}&\textbf{Find}: \{I_{t_i}\}, \{R_{t_i}\}, \{L_{t_i}\}, \{A_{t_i}\}, \forall T = \{v_j, j = 1 : t\},\\ &\text{and } t = |T| \text{ to minimize}: E_{\text{total}}\\ &\text{s.t.}: T_{\text{total}} \leq T_{\text{Deadline}} \text{ and } E_{R_r} \leq E_{\text{limit}}(R_r)\end{aligned}$$

where I_{t_i} is the offloading decision for each task. In particular, $I_{t_i}(= r)$ indicates the task is executed on robot $R_r \in R, \forall r = 1 : n$, and $I_{t_i}(= 0)$ specifies that the task t_i is offloaded to cloud. We assume the total number of robots $n = 3$; hence, the possible task allocation decisions on-board of robot are R_1, R_2, R_3. Then, R_{t_i} is the selection of robot R_r for offloading a task t_i to the cloud, where $R_{t_i} \in I_{t_i} = 0$. This decision is for identifying which robot will offload the task to cloud. Finally, L_{t_i} is the location for each task where the set consists of total l values ($L = 1 \ldots l$), and A_{t_i} is the selected AP for offloaded task where AP set has total α values ($A = 1 \ldots \alpha$). In our formulation, $l = 36$ and $\alpha = 4$.

Experimental Results

We ran extensive simulations on different aspects of our GA scheme for the multi-robot with cloud (GAMRC) approach, and compare it with single cloud-aided robot (GASRC) and GA scheme for a multi-robot on-board (GAMRB) approach where only robot–robot communication has been used for task completion. 40-node taskflow defines tasks that need to be completed in the constrained scenario. The time deadline for the proposed application is 150 s and population size for GA is 500. The stopping criteria for GA is self-maintained and trained to stop running when no changes of results are found for 1500 generations. Among 36 cells in workspace, 9, 18, 20, 33, 36 are considered as obstacle cells. For the multi-robot approach, three robots are considered with processing power of $R1$ (Core i5-4460), $R2$ (Core i5-7600), and $R3$ (Core i5-7600K), respectively. In comparison to the robot processor, the cloud VM is considered to be a minimum M times faster than the fastest robot. For communication modeling, an infrastructure using IEEE 802.11 WLANs is presented with a maximum bit rate of 54 Mbit/s and 8 available stream bit rate (6, 9, 12, 18, 24, 36, 48, 54 Mbit/s). Based on this, the whole workspace is designed with four APs where each AP has three users.

From Table 4.2, the results clearly highlight that GAMRC consumes much lower energy (2836.50 J) than GASRC (4778.80 J) and GAMRB (7057.19 J). In terms of time/delay, GAMRB fails to finish within the time constraint. As for GASRC, even though the tasks are completed within the delay constraint, it is still higher than the fastest GAMRC process. Because of their ability to offload more tasks to the cloud, GASRC and GAMRC entail lower energy. The benefit of GASRC is evident, since the single robot $R1$ allows for the offloading of 21 tasks to the cloud. Despite the fact that only 14 tasks are offloaded for GAMRC, the added trait of local robot–robot communication empowers the robot $R1$ to utilize the nearby robots $R2$ and $R3$ for local completion of tasks or help with offloading tasks to the cloud. This results in better access to resources (cloud VM and local robots) for GAMRC (w.r.t GASRC), more execution of parallel tasks and faster completion of tasks for cloud networked multi-robot systems. Also according to Table 4.2, although the total distance covered by GAMRC (672.8 m) is higher than GASRC (297.98 m), there are three active robots that are moving in tandem to get access to better bandwidth for easier offloading. Besides, the robots are also utilizing their local communication to cover more area effectively. Unfortunately for GAMRB, the lack of cloud availability hampers the system performance. Even the robots cover

Table 4.2 Task offloading performance comparison

Result parameters	GAMRC	GASRC	GAMRB
Generation no.	11,524	8399	N/A
Offloaded task (cloud)	14	21	0
Minimal energy	2836.50 J	4778.80 J	7057.19 J
Total time	80.31 s	109.45 s	413.16 s
Total distance	672.8 m	297.98 m	1161.96 m

Table 4.3 Comparison of robots performance among the three methods (Min: E_{total})

	GAMRC			GASRC	GAMRB		
Robots	R1	R2	R3	R1	R1	R2	R3
Energy constraint	5000 J	1000 J	3000 J	5000 J	5000 J	1000 J	3000 J
Energy consumption	936.18 J	410.78 J	1489.53 J	4778.80 J	2532.2 J	1610.78 J	2914.21 J

the most distance (1161.96 m), it does not provide much benefit as some tasks are too latent and consume high energy in the local processor. In contrast, for GAMRC, the robots offload such tasks to the cloud by moving to the best location, which saves valuable time and energy as reflected in the performance.

Given that each robot has its own energy constraint, proper utilization of energy is a top priority in such applications. Thanks to the involvement of cloud, the energy used for each robot was far within the energy limitation for our GAMRC approach (refer to Table 4.3). By comparison, GASRC process that runs the operation with robot $R1$ incurs an energy of 4778.80 J, significantly higher for a single robot. The same can be said about GAMRB process, which results in a higher value of overall energy as well as higher energy consumption for each of the robot (2532.2 J, 1610.78 J, and 2914.21 J for $R1$, $R2$, and $R3$, respectively). Among these three methods, the lowest energy consumption for the principal robot $R1$ is indeed from GAMRC (936.18 J).

4.3.3 Emergency Management

Through recent developments of IoT and cloud robotics, Industry 4.0 applications could be possibly run with limited human involvement, increased efficiency, and reduced cost [22]. Based on this observation, smart factories with challenges of health, safety, and environment (HSE) are set as our motivation for robotic inspection, maintenance, and repair (IMR). As in the previous setting, a pool of heterogeneous sensors is deployed throughout smart factory to perceive environmental conditions, and cloud-aided robots can further complement this sensing operation with action-oriented task such as inspection, fault diagnosis, etc. All smart factory based applications require robots to continuously update intensive data in order to execute tasks in a coordinated manner.

More specifically, we consider emergency fire management service in smart factory as an illustrative use case scenario. This kind of service requires timely execution to ensure the environmental safety within the factory environment. It consists of multiple tasks such as fire origin and cause identification, human victim and hazardous material detection, evacuation planning, navigation as well as management of external help. Typically, these tasks are interdependent and orchestrated through a directed acyclic graph (DAG) based workflow model. An example workflow for fire emergency management service in the factory is shown

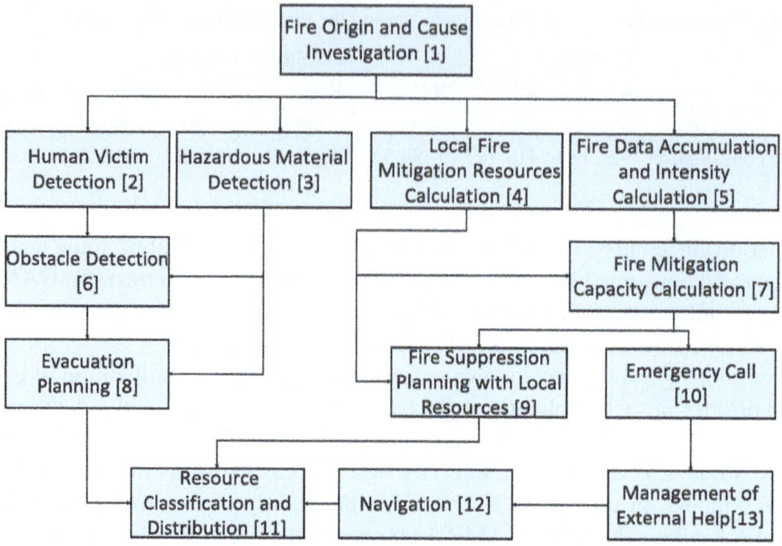

Fig. 4.12 Workflow of fire emergency management service in a smart factory

in Fig. 4.12. The number of tasks in such DAG based workflow may vary from application to application. Here, we focus on allocating computational resources optimally for the tasks of such workflow to jointly minimize the makespan for executing all the tasks, the total energy consumption of resources, and the total cost requirement for executing the tasks on the computational resources.

Obviously, tasks of such emergency management services demand real-time response and data transmission among robots and cloud, because higher inter-communication latency of resources resists the whole system to meet these requirements. To address the limitations of cloud enabled multi-robot systems, the concept of edge cloud infrastructure is introduced between the robot network and cloud as shown in Fig. 4.13. Basically, edge computing is an extension of the cloud computing paradigm, providing data, compute, storage, and application services to end users on a so-called edge layer [11]. It thus assists multi-robot systems by executing the latency sensitive and compute intensive services at closer proximity of the working environment. According to the proposed system model, the edge cloud is aware of the combined resource pool composed of both local and virtual resources to facilitate resource allocation for the tasks.

It is noticed that the total cost of task execution over distributed resources and their uneven energy consumption behaviors significantly affect the system performance. In this case, computational resource allocation appears as a constrained multiobjective optimization problem, when energy consumption of resources, the overall makespan, and total monetary cost for task execution are targeted to minimize simultaneously. Multiobjective evolutionary algorithms are well-known

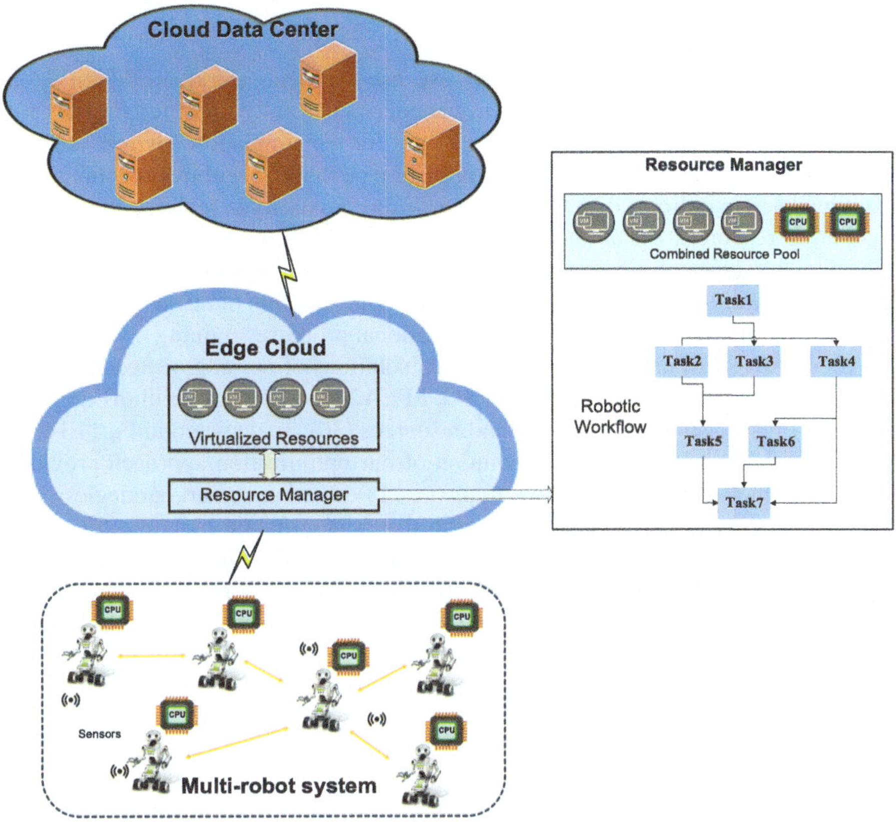

Fig. 4.13 Resource allocation for edge cloud based robotic workflow in smart factory

to generate Pareto-optimal solutions for such multiobjective optimization problems. Therefore, to solve the resource allocation problem, we choose to extend the non-dominated sorting genetic algorithm-II (NSGA-II) [23], capable of finding diverse set of solutions. In more details, we augment the NSGA-II by defining a new chromosome structure, pre-sorted initial population based on the task size and processing speed of the resources. Besides, while crossing over the chromosomes, rather than arbitrary selection, the chromosome having minimum distant solution from the Pareto-front to the origin is selected to balance the values of all objectives in subsequent generations. The results of our simulation experiments significantly improve the existing multiobjective optimization approaches including benchmark NSGA-II, multiobjective particle swarm optimization (MOPSO), strength Pareto evolutionary algorithm II (SPEA2), and Pareto archived evolution strategy (PAES) in terms of meeting the stopping criteria, satisfying data-dependency threshold, and minimizing the total energy consumption, the makespan and the total system cost for varying number of tasks and resources.

Experimental Results

To conduct our experiments in Matlab, we use synthetic workload driven from real-world references and select system parameters carefully on different trials for fair evaluation [24, 25]. The parametric values for the simulation environment are summarized in Table 4.4. The value of simulation parameters within a specific range is set by discrete uniform distribution. Apart from proximity based classification of computing resources (local and virtual), to reflect the resource heterogeneity, we consider three types of resources (low, medium, and high) based on their processing speed for ease of the simulation. Per unit energy consumption of resources vary time to time according to their speed and computational processes running on them.

After 200 generations, the Pareto-optimal solutions of our augmented NSGA-II along with benchmark NSGA-II, MOPSO, SPEA2, and PAES algorithms on fixed number of heterogeneous tasks (50) and resources (30) are depicted in Fig. 4.14. In this scenario, each Pareto-optimal solution of our optimization approach provides better outcome for three objectives, compared to other benchmark strategies. The initial population of the proposed approach that is determined based on the task's size and processing speed of resources inherently minimizes the overall makespan, energy consumption, and total cost. Selection of the chromosome having minimum distant solution from the origin for crossover further improves its performance.

Table 4.4 Simulation parameters

Parameter	Value
Population size	50
No of generations	50–500
Mutation rate	0.5
Crossover rate	0.5
Number of tasks in a workflow	35–65
Number of computing resources	15–45
Processing speed of virtual resources	10,000–30,000 MIPS
Processing speed of local resources	8000–15,000 MIPS
Per unit time (s) energy consumption of virtual resources	50–150 W
Per unit time (s) energy consumption of local resources	40–90 W
Tasks data size	5000–10,000 MI
Per unit time (s) monetary cost of virtual resources	0.50–0.90 Cents
Per unit time (s) monetary cost of local resources	0.25–0.40 Cents
Allowable completion time of all tasks	4000–5000 ms
Maximum allowable energy consumption of workflow	1500–2500 W
Total budget	150–200 Cents
Data-dependency threshold	1200–1500 ms
Communication bandwidth	128–512 Kbps
Ratio of local and virtual resources in resource pool	1/3
Ratio of dependent and independent tasks	2/5

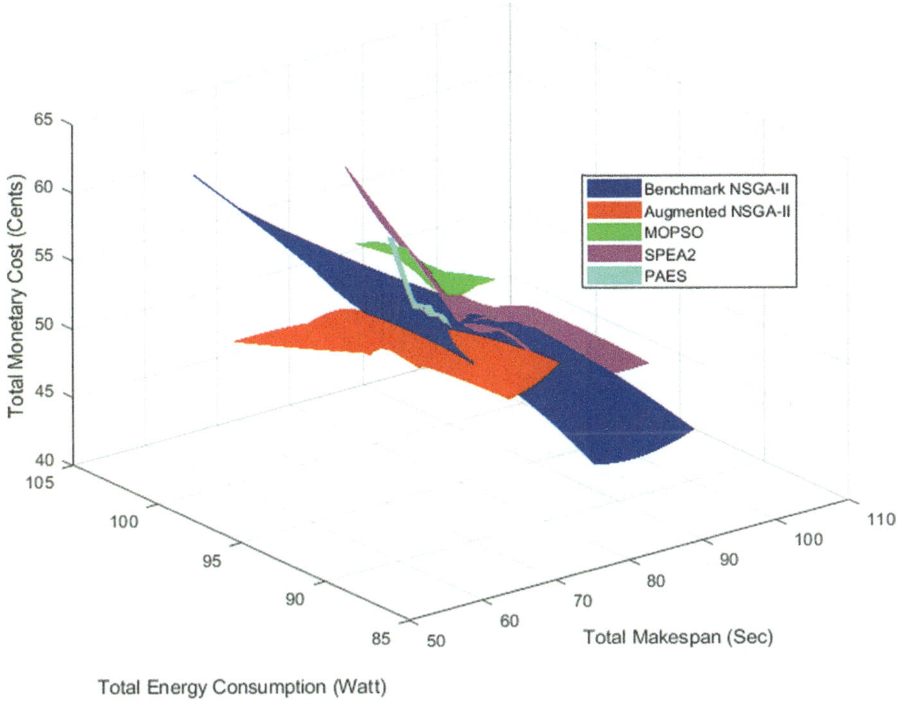

Fig. 4.14 Comparison of Pareto-optimal solutions

Thus, after a fixed number of generations, it provides significantly better solution than the random population used in benchmark NSGA-II. In the selection of chromosome for generating the first child population from parent population, our augmented NSGA-II approach selects the chromosome with the best fitness value. It also complements finding better solutions compared to benchmark NSGA-II, where all chromosomes are considered for applying genetic operators.

In addition, MOPSO follows a one-way information sharing mechanism, and the evolution only looks for the best solution. In comparison to benchmark NSGA-II, MOPSO multiplies the chances to keep individuals' changes, making it easier to maintain diversity in the solution space. However, in terms of Pareto-optimal set quality, our augmented NSGA-II and benchmark NSGA-II algorithm give shorter distinct non-dominated set while generating Pareto-optimal front. Our augmented NSGA-II is relatively more robust as it is less dependent on a random technique. Solutions by SPEA2 have significantly better diversity than NSGA-II ones due to the better diversity maintenance strategy of SPEA2. However, SPEA2 is more computationally expensive to run than NSGA-II and is hence more time consuming. Moreover, in SPEA2 the search populations are randomly initialized, whereas in our augmented NSGA-II a pre-sorted population is initialized heuristically. For these reasons, our augmented NSGA-II gives better Pareto-optimal solutions than SPEA2.

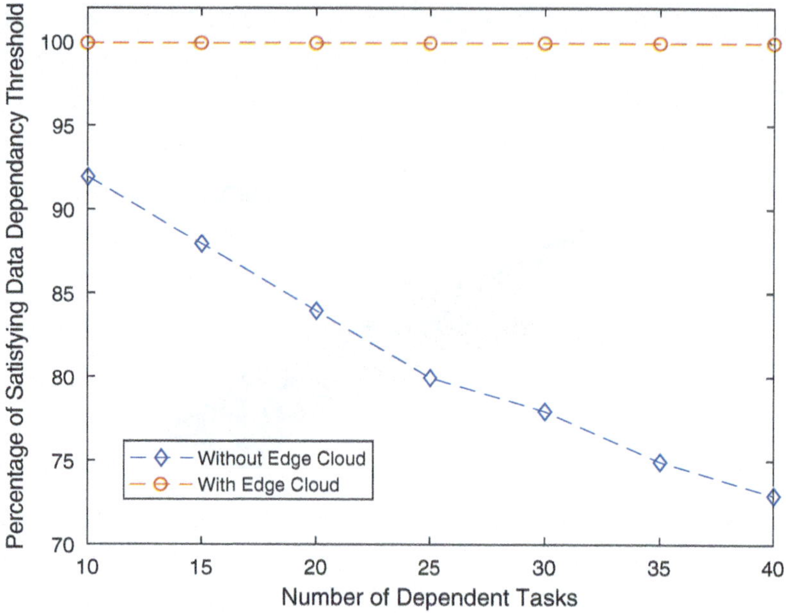

Fig. 4.15 Impact of edge cloud on interdependent tasks

On the other side, PAES uses local search from one current solution to generate a new candidate and concurrently compare the current solution with the candidate solution, and also maintains an archive to keep the best solutions found so far. To maintain the archive, computational complexity of PAES algorithm is the highest among others. This algorithm is also based on random initial population and for larger number of objectives it cannot achieve good results as our augmented NSGA-II and other studied approaches.

On fixed number of generations (200) and computing resources (30), the impact of edge cloud in dealing with varying number of interdependent tasks of robotic workflow is represented in Fig. 4.15. Compared to cloud based multi-robot system, the edge cloud based system performs significantly better. Since, edge cloud brings virtual computing resources closer to the robots, distributed placement of interdependent tasks on both local and virtual resources does not violate the data-dependency threshold of the dependent tasks. For validating the impact of edge cloud on interdependent tasks, we use the percentage of satisfying data-dependency threshold $\hat{\delta}_T$ as performance metric. It is calculated by the ratio of number of tasks of a workflow that maintain the data-dependency threshold s_T and the total number of dependent tasks on that workflow d_T as Eq. (4.1). The higher percentage denotes higher efficiency of the system to handle interdependent tasks of a robotic workflow.

$$\hat{\delta}_T = \frac{|s_T|}{|d_T|} \times 100\% \tag{4.1}$$

Figure 4.15 depicts that for increasing number of dependent tasks, the percentage of satisfying data-dependency threshold remains constant with almost zero violation. As the cloud is deployed geographically far from the edge cloud, with the increasing number of dependent tasks, the violation of data-dependency threshold increases due to higher communication latency.

4.4 Multi-Robot Fleet Formation

In some applications, there is a need for multiple mobile robots to collaborate on a task. A fleet of robots in linear formation is the most common form of collaboration which has a wide range of applications. For example, in the robot transport scenario, the navigator has fully equipped sensors, capable of navigation and avoidance, while the following robots in the convoy are designed for low cost and heavy load. Without navigation, they can only track the pilot's trajectory. Another example is that when a mobile robot loses its navigation ability due to a fault, degenerating into the following mode, it will need the guidance of a normal robot backing to the robot port.

Most robot formation and follow methods rely on the global positioning information of the robots, which limits the scope of use, since it is difficult to obtain the global position in a strange or narrow environment.

Therefore, this section introduces a formation-control system scheme, making use of speed information of the leading robot, and visual recognition results. In this scheme, the relative position of leader and follower can be obtained from visual recognition and leader speed information sharing; thus, the dependence of global positioning information is eliminated. However, there comes inevitable challenges, including the real-time transmission of a large amount of data, and computing power requirements for image processing in the whole system. These issues can be overcome by introducing the concept of fog computing and providing fog network domain with general computing capacity from FNs. With fog-enabled SLAM, the leading robot firstly builds a map with the help of FNs, and then uses the map to lead its followers.

4.4.1 System Architecture

The architecture of robot fleet formation is shown in Fig. 4.16, which is composed of the following components.

1. **Mobile Robot**
   The robots are based on 4-wheel differential driving platforms, and all the mobile robots are equipped with sensors including inertial measurement unit (IMU), wheel encoder sensor. Raspberry pi 3b and Stm32 MCU on the robots are used

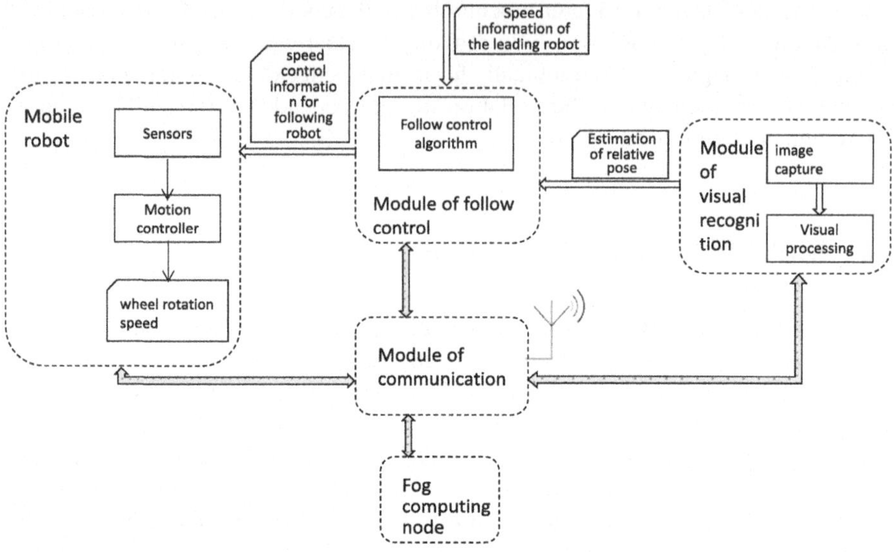

Fig. 4.16 Architecture of the fog-enabled formation-control system

for calculation and control, respectively. Inputting desired linear velocity and angular velocity to the motion controller, it will convert them into the motion of robot, with PID control of wheel rotation. With a LIDAR sensor, the leader of the fleet acts as a leader, it can navigate automatically.

2. **Module of Visual Recognition**
A camera is mounted on the central axis of each robot, and it continuously takes pictures of the robots in front of it. A visual processing algorithm is used to extract the feature points of the leading vehicle and estimate its relative distance and pose, which is then going to be used in follow control algorithm.
3. **Module of Follow Control**
The main function of this module is to receive two inputs, the speed information of the robot in front, the result of visual processing module, and then figures out the speed control information of the following robot.
4. **Fog Computing Node**
In the context of fog computing paradigm, a FN refers in particular to the device that has general computing, storage, and communication capabilities. The form of device could be a router, a server, or even a smart phone. The role of the FN is to provide task offloading for application in the network, for the consideration of energy saving, performance improvement, etc.
5. **Module of Communication**
The communication module is responsible for information sharing and interaction between different components. The communication module establishes a local network domain. The robots are connected to the network via wireless, while the fog computing nodes connect to the network by Ethernet.

The system is based on ROS, which is a widely used open-source software platform for robots. Thanks to its distributed feature, components of a service can be run on any desired nodes. For example, module of visual processing requires a lot of computing resources, so it is implemented on FN, while module of follow control interacts directly with onboard MCU, and it is better run on raspberry pi computer which is carried on the robot.

4.4.2 *Fog-Enabled Solution*

In the above architecture, the fog computing service can be sold in market. This can encourage owners of free computer to share the computing power and receive rewards. Both the computing power and application service can be sold on a blockchain-based platform. End users and fog computing service providers can complete the transactions online in a manner of "Pay-As-You-Go." Here we use iExec as an example blockchain platform for fog computing service transactions.

iExec is a blockchain-based decentralized cloud platform which makes use of idle computing resources to handle delicate tasks, while the result of execution is verified by PoCo, a smart contract on chain. Actually, there are three participation roles in iExec, end user, resource provider, and Dapp provider. With the help of iExec, the robot-rescue service can be shared as a Dapp deployed on its market. End users may easily access the service by ordering the corresponding Dapp, and pay for it in electronic currency. The components of iExce platform are as shown below, which mainly consists of on-chain part and off-chain part shown in Fig. 4.17.

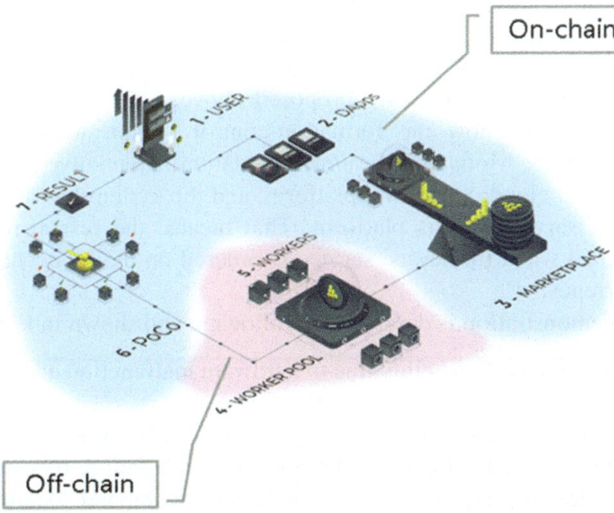

Fig. 4.17 The components of iExce platform

- **DApp**:
  The application that can be executed on iExec in the form of Docker image;
- **Marketplace**:
  An online market lists available computing resources with each attributes, including quantity, reputation, price, etc. Users are able to order a certain Dapp on marketplace;
- **Worker Pool**:
  The worker pool is composed of multiple workers, and its manager is responsible for pricing and quoting in the market;
- **Worker**:
  A worker usually joins a resource pool, and this is where the Dapp is actually executed;
- **PoCo**:
  An on-chain program to verify the credibility of the computing results.

The idea putting robot-rescue service to iExec market is to prove that those vertical industry applications, especially in smart cities scenarios, can be organized through a blockchain-based platform, benefiting both to provider and end user. To this end, a script that can trigger local service is firstly created and is packaged into Docker image. We push this image to Docker Hub and register to iExec platform. From now on, it can be ordered by users, known as a Dapp. Once an end user decides to use this service, he can order this Dapp via iExec marketplace, specifying custom parameters. Then the rest of the work is atomically done by iExec platform—the task is send to the worker pool scheduler, and eventually assigned to a worker. When the worker receives the task, it pulls the Dapp image from Docker hub, checking the hash value before running it. The started Docker container then sends command to robot runtime environment and triggers the rescue service. Next, as you would expect, the rescuer is sent out to explore the space, generating a map and performing robot formation. The described work flow of service provider and end user is illustrated in Fig. 4.18.

Based on the system architecture proposed above, a demonstration is implemented to demonstrate how the formation-control system works on rescuing malfunctioning robots. Moreover, the implementation has involved a blockchain-based decentralized cloud computing platform, and the system has also connected the robot-rescue service to this platform. That means, the rescue robot can be remotely launched when the rescue service is ordered on the blockchain platform with digital currency.

The whole demonstration is described as following and shown in Fig. 4.19.

1. Two robots lost navigation ability due to hardware malfunction and were trapped in Zone A.
2. Fortunately, robot-rescue services are available nearby. With the blockchain-based distributed cloud computing platform, owner of failure robots can choose and order a desired rescue service, specifying the trapped location and safe destination.

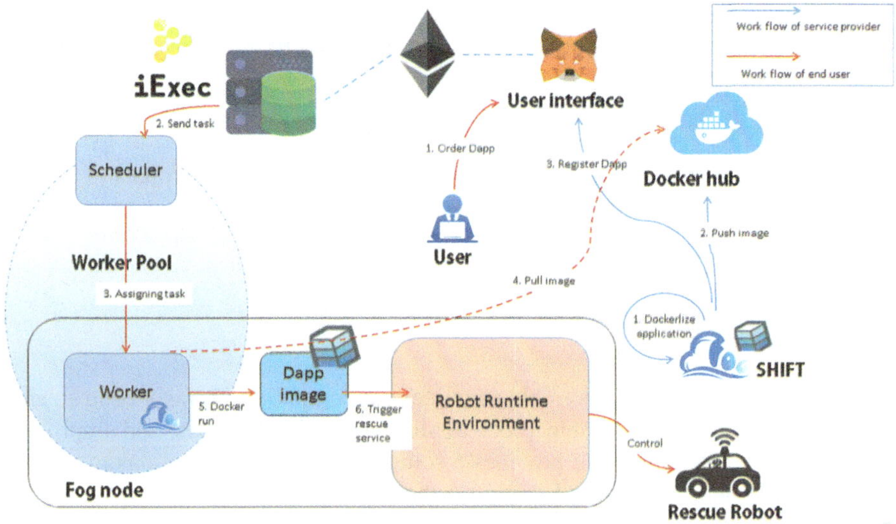

Fig. 4.18 Work flow of service provider and end use

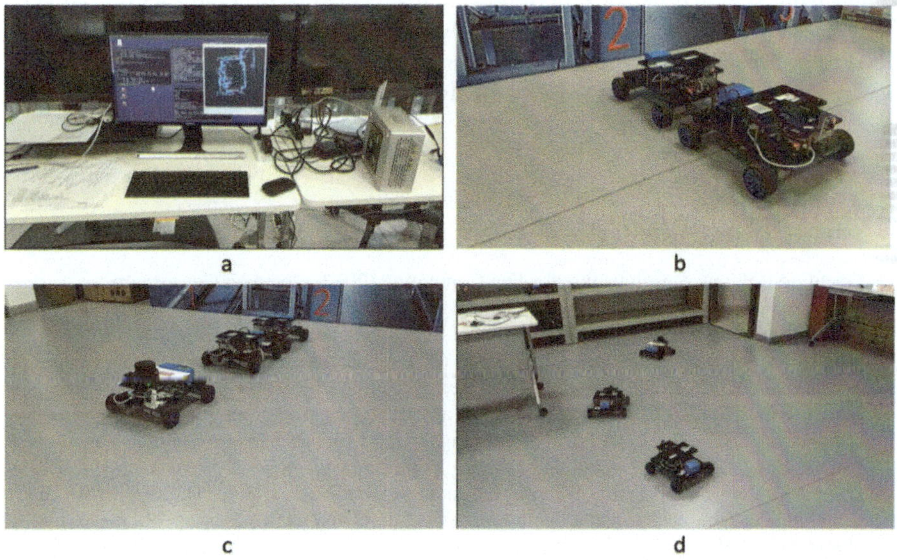

Fig. 4.19 (**a**) Rescue service is connected to blockchain-based platform. (**b**) Two trapped robots. (**c**) The rescue robot approaches two trapped robots. (**d**) The robot fleet moving to destination

3. The rescue robot is firstly sent out to explore the environment of trapped zone. With support of fog computing, it builds a map, and then it plans a route to the destination.
4. After that the rescue robot approaches the trapped robots, and requests them to follow.
5. The rescue robot acts as a leader, along with other two trapped robots to form a fleet, heading towards the destination.

4.5 Conclusion

Fog computing employs all possible resources for computing, storage, control, and networking, and then to enable robot applications with lost cost and improved efficiency. The fog computing provides a solution for robot applications to realize the perception and autonomy of things on the edge of distributed information processing. However, when designing and implementing a robot system, manufacturers should consider how the physical technologies can work best for them. As such kind of system requires to deal with massive amount of data from different components in different formats, efficient data analytics and machine learning model is supposed to enhance the system efficiency. In future, the fixed network structures will very likely be replaced with dynamic network and edge intelligence. The researchers therefore need to focus on resource optimization as well to reduce the energy consumption and system development cost by effectively leveraging fog computing technologies.

References

1. OpenFog Reference Architecture for Fog Computing (2017) OpenFog Consortium
2. Dastjerdi AV, Gupta H, Calheiros RN, Ghosh SK, Buyya R (2016) Fog computing: principles, architectures, and applications. In: Internet of things. Elsevier, Amsterdam, pp 61–75
3. Hou X, Li Y, Chen M, Wu D, Jin D, Chen S (2016) Vehicular fog computing: a viewpoint of vehicles as the infrastructures. IEEE Trans Veh Technol 65(6):3860–3873
4. Dey S, Mukherjee A (2016) Robotic slam: a review from fog computing and mobile edge computing perspective. In: Adjunct proceedings of the 13th international conference on mobile and ubiquitous systems: computing networking and services. ACM, New York, pp 153–158
5. Cadena C, Carlone L, Carrillo H, Latif Y, Scaramuzza D, Neira J, Reid ID, Leonard JJ (2016) Simultaneous localization and mapping: present, future, and the robust-perception age. CoRR abs/1606.05830
6. Wang J, Xu J, Yang Y, Xu H (2017) GPP based open cellular network towards 5G. China Commun. 14(6):189–198
7. Dinh TQ, Tang J, La QD, Quek TQ (2017) Offloading in mobile edge computing: task allocation and computational frequency scaling. IEEE Trans Commun 65(8):3571–3584
8. Hu G, Tay WP, Wen Y (2012) Cloud robotics: architecture, challenges and applications. IEEE Netw. 26(3):21–28
9. Saha O, Dasgupta P (2018) A comprehensive survey of recent trends in cloud robotics architectures and applications. Robotics 7(3):47

10. Yan H, Hua Q, Wang Y, Wei W, Imran M (2017) Cloud robotics in smart manufacturing environments: challenges and countermeasures. Comput Electr Eng 63:56–65
11. Afrin M, Jin J, Rahman A, Tian YC, Kulkarni A (2019) Multi-objective resource allocation for edge cloud based robotic workflow in smart factory. Futur Gener Comput Syst 97:119–130
12. Afrin M, Jin J, Rahman A (2018) Energy-delay co-optimization of resource allocation for robotic services in cloudlet infrastructure. In: International conference on service-oriented computing. Springer, Berlin, pp 295–303
13. Liu J, Xu W, Zhang J, Zhou Z, Pham DT (2016) Industrial cloud robotics towards sustainable manufacturing. In: ASME 2016 11th international manufacturing science and engineering conference. American Society of Mechanical Engineers, New York
14. Rahman A, Jin J, Cricenti AL, Rahman A, Kulkarni A (2018) Communication-aware cloud robotic task offloading with on-demand mobility for smart factory maintenance. IEEE Trans Ind Inf 15(5):2500–2511
15. Lasi H, Fettke P, Kemper HG, Feld T, Hoffmann M (2014) Industry 4.0. Bus Inf Syst Eng 6(4):239–242
16. Wan J, Chen B, Imran M, Tao F, Li D, Liu C, Ahmad S (2018) Toward dynamic resources management for IoT-based manufacturing. IEEE Commun Mag 56(2):52–59
17. Wan J, Chen B, Wang S, Xia M, Li D, Liu C (2018) Fog computing for energy-aware load balancing and scheduling in smart factory. IEEE Trans Ind Inf 14(10):4548–4556
18. Mahmud R, Ramamohanarao K, Buyya R (2018) Latency-aware application module management for fog computing environments. ACM Trans Internet Technol 19(1):9
19. Bonomi F, Milito R, Zhu J, Addepalli S (2012) Fog computing and its role in the internet of things. In: Proceedings of the first edition of the MCC workshop on Mobile cloud computing. ACM, New York, pp 13–16
20. Bonkenburg T (2016) Robotics in logistics: a DPDHL perspective on implications and use cases for the logistics industry. In: DHL customer solutions & innovation, Bonn, p 4
21. Rahman A, Jin J, Rahman A, Cricenti A, Afrin M, Dong YN (2019) Energy-efficient optimal task offloading in cloud networked multi-robot systems. Comput Netw 160:11–32
22. Zhong RY, Xu X, Klotz E, Newman ST (2017) Intelligent manufacturing in the context of industry 4.0: a review. Engineering 3(5):616–630
23. Deb K, Pratap A, Agarwal S, Meyarivan TA (2002) A fast and elitist multiobjective genetic algorithm: NSGA-II. IEEE Trans Evol Comput 6(2):182–197
24. Wang L, Liu M, Meng MQ-H (2016) A pricing mechanism for task oriented resource allocation in cloud robotics. In: Robots and sensor clouds. Springer, Berlin, pp 3–31
25. Rahman A, Jin J, Cricenti A, Rahman A, Panda M (2017) Motion and connectivity aware offloading in cloud robotics via genetic algorithm. In: IEEE global communications conference, pp 1–6

Chapter 5
Fog-Enabled Wireless Communication Networks

Wireless networks are and will continue to be a fundamental cornerstone of the global digital economy and our connected society, provisioning diverse services for people and things. The proliferation of smart user devices, such as smart phones, laptops, and tablets, is pushing the current wireless networks to their limits. Wireless networks are experiencing an unprecedented traffic growth with an estimated compound annual growth rate of 0.6–1.0 and an increasing variety of services and applications, each with potentially different traffic patterns and quality of service (QoS) and quality of experience (QoE) requirements, for example, ultra-high data rate and/or reliability, and ultra-low latency. To cope with the continuing traffic growth and service expanding, future wireless networks will have to be heterogeneous and densely deployed, featuring the coexistence of different radio access technologies (RATs), such as LTE/LTE-advanced, Wi-Fi, IoT, 5G NR, etc.

Accordingly, future wireless networks will be significantly more complex to deploy and operate than the existing 3G/4G mobile networks, due to the dense deployment of small BSs and the heterogeneities of network nodes, RATs, and services (and hence traffic patterns). The increasing management complexity of wireless networks has made their self-optimization a necessity, where wireless networks are automated to minimize human intervention and to proactively optimize network deployment, operation, and multi-RAT resource allocation to meet increasing service demand from people and things.

Historically, 3GPP Release 8 first introduced the concept of self-organizing networks (SONs), aiming to automate the operation and management of wireless networks. SON functionalities can be generally classified into self-configuration, self-healing, and self-optimization. In subsequent 3GPP releases, SON concepts and technologies have been extensively developed. Nowadays, SON technologies are widely expected to: (1) enhance the intelligence and autonomous adaptability of wireless networks, especially the radio access networks (RANs), (2) reduce the capital and operational expenses (CAPEX and OPEX) for wireless network operators, and (3) improve both network-wide performance and user specific QoS

Y. Yang et al., *Fog-Enabled Intelligent IoT Systems*,
https://doi.org/10.1007/978-3-030-23185-9_5

and QoE [1]. Moreover, future wireless networks will require (near) real-time and scalable SON solutions that are tailored for multi-vendor and multiple RAT networks, as well as for machine-to-machine (M2M) communications and IoT systems.

Wireless network self-optimization comprises various mechanisms that optimize network parameters during operation according to measurement data taken at different parts of the network. The number of network parameters that need to be self-optimized are still increasing and would be enormous in the near future. Nevertheless, most existing SON solutions are mainly based on heuristics, with the automated information processing limited to relatively simple methods [2]. Many open challenges of SON remain unsolved, for example, the automated coordination between different SON functions, and the trade-off between centralized and distributed SON implementations.

In 2012, fog computing was proposed by Cisco as an extension of the centralized cloud computing paradigm [3]. Since then, fog computing has been considered as a promising paradigm shift to enable autonomous management and operation of wireless networks. Fog computing features a distributed computing infrastructure, its proximity to the network edge and end users, and the dense geographical distribution of fog nodes (FNs). These allow fog computing to exploit the local signal processing and computing, cooperative resource management, and distributed storing/caching capabilities at the network edge [3].

A FN is typically a virtualized platform hosted on either a dedicated computing node, which is equipped with communication interface(s), or a networking node/device, such as a BS, an AP, a router, or a switch. This asks for the combination of fog computing and two other emerging technologies: software defined networking (SDN) and network function virtualization (NFV). SDN implies a logically centralized network control plane, which allows the implementation of sophisticated mechanisms for traffic control and resource management. In SDN, the control plane carries signaling traffic, calculates routes for data flows, and performs configuration and management for the network, while the data plane is responsible for transporting data packets [4]. As defined by the ETSI, NFV implements network functions in software that can run on a range of industry standard server hardware and that can be moved to or instantiated in various locations of the network as required, without the need to install new equipment. Fog computing, in conjunction with SDN and NFV, can bring extensive programmability and flexibility into wireless networks, and thus enable distributed and intelligent SON functionalities in (near) real-time.

In a fog-enabled wireless network, a large amount of signal processing and computing is performed in a distributed manner, and local data can be stored and processed in edge devices, such as BSs, APs, and user devices, thus enabling applications that require very low and/or predictable latency and offering mobility support, geo-distribution, location awareness, and low latency [5]. For instance, mobile application processing delays can be reduced by offloading the associated computationally intensive tasks to the FNs that are close to the corresponding mobile applications.

Overall, the fog computing paradigm will allow the self-optimized computing, control, caching, storage, and networking functions to be dynamically relocated among the cloud, the fog, the network edge, and the things, as well as allow self-optimized management of network function and service lifecycles. Hence, fog computing will lead to new opportunities in the design of SON for wireless networks, by exploring the various trade-offs between distributed and centralized operation, between local and global optimization, etc.

In this chapter, we explore fog-computing enabled self-optimization for wireless networks, which will act as the infrastructure to provision ubiquitous wireless connectivity for the IoT. More specifically, we will first discuss different SON architectures and how they would benefit from the fog computing paradigm, and then look into how fog computing would provide new opportunities and enable new features for several important SON functionalities, including mobility load balancing, self-optimization of mobility robustness and handover, self-coordination of inter-cell interference, self-optimization of coverage and capacity, and self-optimized allocation of computing, storage, and networking resources in wireless networks.

5.1 Introduction

Different network architectures have been considered for the realization of SON, including centralized SON, distributed SON, and hybrid SON. The choice of the SON architecture may affect the performance and/or efficiency of the SON functionalities for wireless networks.

5.1.1 Centralized SON

In centralized SON (C-SON), the self-optimizing algorithms mainly reside in the network management system (NMS) or in the operation and maintenance center (OMC) [1], relying on the collection of information from all involved network nodes to identify and tackle network-wide problems. It is intuitive to build C-SON on a cloud radio access network (CRAN), which centralizes baseband processing in a pool of baseband units (BBUs) centralized in a cloud server by decoupling the baseband processing from the radio frequency (RF) transmissions of remote radio heads (RRHs) [6]. C-SON may thus benefit from the available global information about network status and key performance indicators (KPIs), as well as the powerful computational capability at the cloud server to run self-optimizing algorithms for a potentially large-scale wireless network. Since the self-optimizing algorithms and network control are centralized, C-SON solutions are relatively robust against potential network instabilities caused by the parallel operation of multiple SON functions with conflicting objectives.

However, the scalability of network monitoring and management remains a big challenge to any centralized network architecture. With all processing and control being centralized in a CRAN, the cloud server could be overloaded, and the user experience of latency-sensitive applications may be affected by the potentially large delays. Consequently, in heterogeneous networks (HetNets) with a dense deployment of small cells, CRAN-based C-SON solutions may not be able to respond quickly enough to the transient traffic demand from dense small cells.

The scalability and flexibility of C-SON solutions can be improved by exploiting fog computing as a complement to cloud computing [5]. In a fog-computing enabled RAN, fog computing can be combined with software-defined wireless networking (SDWN), where the control plane and the user data plane are decoupled in FNs or in a FN gateway. Each BS is connected to a nearby FN, which provides computing, processing, and storage capabilities to all the associated BSs and mobile devices in its coverage area, while the FNs are well connected with each other and to a central cloud server [3].

SDWN provides application programming interfaces (APIs) to facilitate the programmability of network operations. The APIs define the interactions between a higher-layer controller and a lower-layer controller [7]. A higher-layer controller defines specific policies and sends these policies to its associated lower-layer controllers to instruct them how specific applications or services should be processed based on their local information. The control-plane functions can be programmed directly in the controllers of different hierarchical layers.

The fog-enabled SDWN can construct a hierarchical control plane [7, 8], where a higher layer of controllers are deployed in the FN gateway, a lower layer of controllers are deployed in the FNs, and the different controllers need to cooperate for accomplishing specific networking functionalities [9]. If needed, more hierarchical layers of logical controllers can be deployed in the cloud server, the network edge nodes, and even user devices. A higher-layer controller manages a number of lower-layer controllers. Specifically, a single control plane deployed in a FN can make C-SON decisions for all the associated BSs and APs. The logically centralized SON functions in the FNs (or in the FN gateway) can avoid the increase in control signaling overhead even for a large number of distributed heterogeneous network nodes [10]. Moreover, with the combination of fog computing and SDN and NFV technologies, C-SON functions can be virtualized and run on general purpose hardware, thus largely improving the scalability of C-SON.

5.1.2 Distributed SON

In distributed SON (D-SON), the SON functions are distributed across the edge of the network, typically in BSs and APs. Although D-SON algorithms run separately in corresponding network edge nodes, they can directly exchange necessary information with each other in a local area, thus enabling the RAN to adapt to local changes swiftly. Each BS or AP can initiate their SON functionalities and make self-optimization decisions independently or in coordination with neighboring network nodes. Moreover, D-SON solutions are designed for offering near real-time response, i.e., in seconds or even milliseconds, as required by small cells and dense HetNets [1]. Therefore, the D-SON architecture is more flexible than the C-SON architecture.

It is worth noting that the design and deployment of D-SON functions need to consider practical aspects such as response time, computational complexity, size of data sets, and the computational and storage capabilities of individual network edge nodes. Meanwhile, since D-SON functionalities are typically designed as stand-alone functions, D-SON solutions are likely to be vulnerable to network instabilities caused by the parallel execution of multiple SON functions with conflicting objectives. The distributed intelligence embedded in fog computing can be exploited to support D-SON functions for frequent services and real-time applications [7]. In a fog-computing enabled RAN, a local SON coordinator can be deployed at a FN to coordinate the D-SON functionalities distributed across the network edge nodes in the local area. When stand-alone D-SON functions are executed concurrently in the same network edge node or in neighboring edge nodes, the nearby FN will act as the local SON coordinator to avoid possible oscillations of parameter settings caused by two or more SON functions trying to optimize a same set of output parameters but for different objectives.

5.1.3 Hybrid SON

Hybrid SON (H-SON) combines the centralized network management and the distributed SON functionalities. H-SON would be especially desirable for future user-centric, service-aware, and/or content-centric wireless networks, where user devices can be collectively served by all the nearby FNs and their associated small-cell BSs [11]. Fog computing, in conjunction with cloud computing, provides the opportunity to optimize the deployment locations for SON functions and to find a good trade-off between the centralized and distributed SON implementations.

In a fog-computing enabled H-SON architecture, the centralized and distributed SON functions are located in the cloud server and the FNs, respectively. Meanwhile, a FN can also run some simple centralized SON functions to automate the control of its associated network edge nodes and user devices. With fog computing being combined with NFV and SDN, the logically centralized SDN controller can maintain a global view of the wireless network, control data-plane devices, and provide a programming interface for network management applications [4]. The communications between the SDN controller and the data plane devices are typically based on the OpenFlow protocol [12], where the signaling latency between the SDN controller and the SON functions will be kept very low. The SON functions running in FNs can be coordinated and/or managed by the centralized SON functions in the cloud server.

In addition to enabling H-SON, fog computing can also reduce the traffic/signaling burden on the fronthaul and backhaul and the design and computational complexities of centralized SON functions in the cloud server. This is because the FNs are able to undertake most data processing tasks locally, and forward only necessary information to the cloud server, thus reducing the use of network resources and the traffic burden on the fronthaul and the backhaul. Such a fog-computing enabled H-SON architecture simplifies the design, configuration, and management of SON functions, and improves the scalability of self-optimized wireless networks.

5.2 Self-Optimization of Mobility Management

Future heterogeneous and dense wireless networks will feature the coexistence of different RATs, including LTE/LTE-advanced, Wi-Fi, IoT, 5G NR, etc. Accordingly, mobility management in wireless networks is evolving from handling simple single-RAT handovers to managing complicated multi-RAT mobility events. The mobility management complexity will be further increased by the envisioned (ultra-) high density of network nodes in conjunction with the high mobility and high density of user devices.

5.2.1 *Mobility Load Balancing*

Mobility load balancing (MLB) is a self-optimization function that aims to maximize the network capacity (or to optimize user QoS or QoE) through optimizing the distribution of user traffic load across the network nodes and radio resources. As a result, traffic overload or congestion at any network node can be avoided. MLB is usually implemented in a distributed manner and relies on the traffic load estimation and radio resource utilization status exchanged among neighboring network nodes via the X2 interface [13].

Fog computing can be used to enhance the operations, administration, and maintenance (OAM) processes for distributed MLB in support of multi-RAT coordination [14]. To enable unified multi-RAT access and seamless mobility in future wireless networks, a FN can be used to provide centralized management for the associated multi-RAT BSs and/or APs in a local area. In a fog-computing enabled RAN, a FN can timely collect and distribute the location information of mobile devices of different RATs. Thus, useful information for MLB can be extracted by collecting, classifying, and analyzing data streams in the network edge. The fog-embedded SDN controller can reactively or proactively instruct the data plane devices to identify and handle different traffic flows in the network appropriately [12]. Local monitoring data of cells with a high handover rate can be collected and processed at a nearby FN. Furthermore, to achieve autonomous

MLB, reinforcement learning, e.g., distributed Q-learning, can be implemented at FNs to learn for each traffic load state the best MLB action to take while maintaining the required handover performance [15].

5.2.2 Mobility Management and Procedures

In mobile communications, handover management is one of the most critical techniques to guarantee the QoS requirements of mobile users and to provision seamless user experience. In a heterogeneous network (HetNet), high speed user devices should be served by macrocells with large coverage areas and reliable connections, while low mobility user devices should be served by small cells that can provide a very high capacity to a small number of user devices in a small coverage area [11]. In multi-tier HetNets, unnecessary handovers (e.g., ping-pong handovers) or handover caused radio link failures are more likely to happen as compared with conventional single-tier cellular networks, due to the smaller coverage areas of small cells and severer inter-cell interference [16, 17]. Moreover, frequent handovers in dense HetNets also cause a heavy overhead on the fronthaul, the backhaul, and the core network [11, 18].

Mobility robustness/handover optimization (MRO) is a self-optimization function that minimizes the number of call drops, radio link failures, and unnecessary handovers (including ping-pong handovers) by optimizing the handover parameter settings (e.g., the time to trigger) or the cell reselection parameter settings for a mobile device in the connected mode or in the idle mode, respectively. Similar to MLB, MRO is typically implemented in a distributed manner. MRO related control messages include S1-AP or X2-AP handover request, handover report, and radio link failure indication message. To realize MRO, each mobile device needs to register its location when it first connects to the network, and will need to report its updated location information to the network periodically. A smart mobile device should be able to identify its geographic location by employing one of the existing wireless localization technologies, such as Wi-Fi based localization, LTE mobility management, and Bluetooth low energy beacon based localization [8].

Fog computing enables scalable and on-demand mobility management, by exploiting fog computing capabilities for mobile device location registration and for handover management. In a fog-computing enabled wireless network, the home subscriber servers can be distributed in the FNs, so as to shorten location registration delays and to reduce the traffic load on the backhaul by making the home subscriber servers closer to mobile devices. To avoid frequent handovers and to alleviate the control overhead of handovers, many handover related functions such as handover and admission control can be performed at FNs and their associated BSs or APs, in conjunction with appropriate user-cell association mechanisms. Moreover, macrocell BSs, small-cell BSs, and FNs can shift some handover related control and decision-making to mobile devices [11].

In a fog-computing enabled RAN, the handover decision for a handover between two FNs is made between the source FN and the fog gateway. The event of a handover happening on the border between two FNs may occur in two cases: (1) an active user device, who has initialized a session outside the source FN coverage area, moves across the source FN coverage area; and (2) a user device, who initializes a session within the coverage of the source FN, remains in the active state while moving out of the source FN coverage area. If the fog-enabled RAN architecture deploys S1 and X2 interfaces, then the source and target FNs can communicate with each other through the S1 interface. The handover signaling flow between the user device, the source FN, and the target FN via the fog gateway is shown in Fig. 5.1 (see Fig. 3 in [11]), where the signaling in layers 1–3 and the transmissions of user data can be divided into three stages: handover preparation, handover execution, and handover completion.

The handover signaling overhead is composed of the processing overhead and the transmitting overhead. The processing overhead consists of the signal processing costs at the user device, the FN, the fog gateway, and the core network. The transmitting overhead includes the transmission associated costs between the source FN and the fog gateway, and between the fog gateway and the core network.

In a fog-enabled SDWN, where the control plane and the user data plane are decoupled in FNs or in a FN gateway. The control plane deployed in a FN (or in the FN gateway) can support centralized control-plane handover decisions for all the associated BSs and APs. This can avoid the increase in control signaling overhead due to frequent handovers in dense HetNets [10].

After a handover has been executed successfully, the data for the mobile user will be transmitted through the target FN and the target BS/AP to the user. In order to simplify multi-RAT cooperation in future HetNets, only IP protocols will be used to support signaling interactions in the control plane. Existing interfaces can be made open so that a unified interface protocol can operate flexibly. In complicated handover scenarios, the hierarchical SDWN controllers located in the cloud server, the FN gateway, the FNs, and the network edge nodes/devices will need to cooperatively manage the mobility and complete the handover signaling procedures.

The performance of the fog-computing enabled handover procedure was compared with that in a conventional RAN in terms of system signaling overhead through computer simulation in [11], where the signaling overhead was assumed to be proportional to the delay required to send or process a signaling message and had no unit. In the simulation, the session holding times were generated as random variables following independent and identical exponential distributions. The session arrivals were generated following a Poisson process with an average arrival rate of λ (in number of sessions per minute). Figure 5.2 (see Fig. 4 in [11]) plots the signaling overhead of the fog-computing enabled handover procedure (denoted by "FRAN") and that in a conventional RAN (denoted by "non-FRAN") versus the average session arrival rate λ for handovers between two femtocells in non-FRAN with or without control-user-plane split, handovers between two FNs (referred to as "F-AP" in Fig. 5.2) in FRAN, handovers between a femtocell and a macrocell in non-FRAN

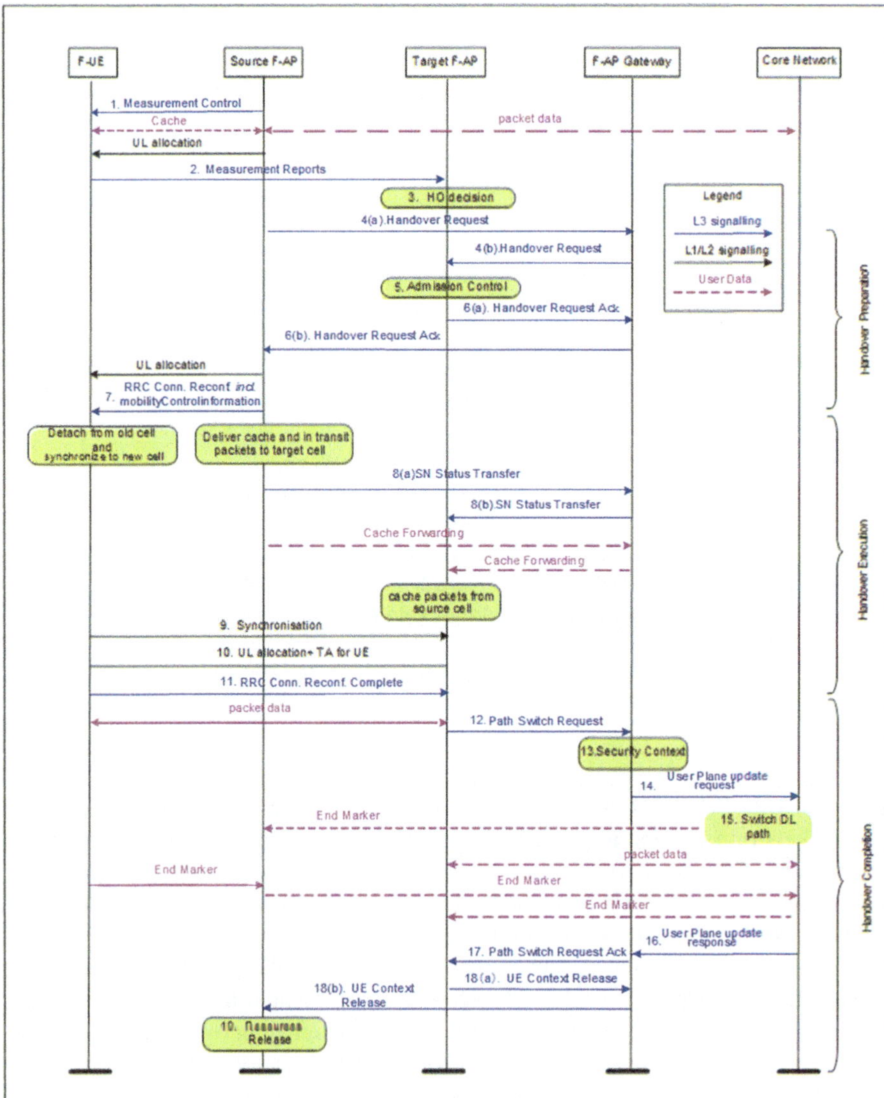

Fig. 5.1 The handover signaling flow between the user device (denoted by "F-UE"), the source FN (denoted by "Source F-AP"), and the target FN (denoted by "Target F-AP") via the fog gateway (denoted by "F-AP gateway") [11]

with or without control-user-plane split, and handovers between a FN and a macro-RRH (MRRH) in FRAN. We can see that the fog-computing enabled handover procedure results in a signaling overhead much lower than that of a conventional RAN handover procedure. This is because the fog-computing enabled RAN takes full advantage of the computing, processing, and caching capabilities in the FNs,

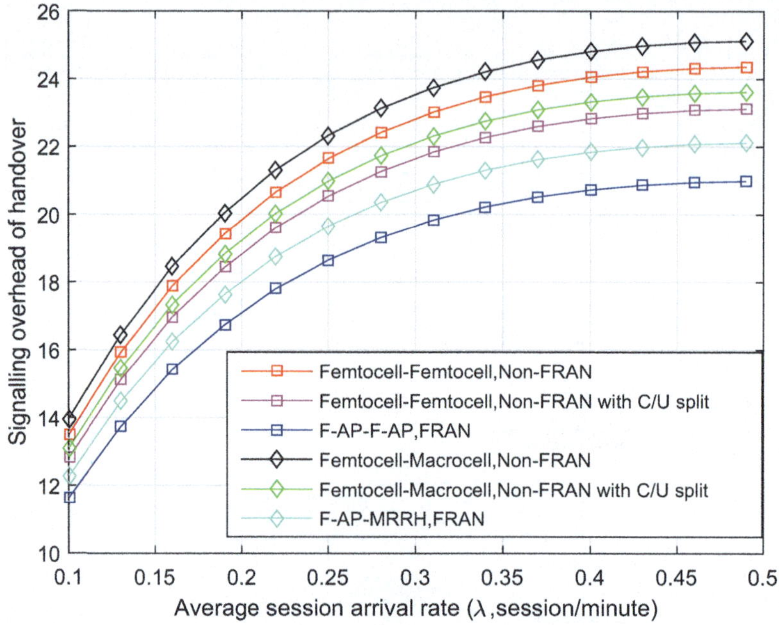

Fig. 5.2 Signaling overhead of the fog-computing enabled handover procedure (denoted by "FRAN") and that in a conventional RAN (denoted by "non-FRAN") versus the average session arrival rate [11]

small-cell BSs, APs, and user devices, thus avoiding the transmission of the entire data packets or a large amount of control signaling to a distant centralized server, e.g., a BBU pool. At the same time, the handover decisions are made between FNs and the fog gateway rather than in the BBU pool, leading to a significant reduction in the transmitting overhead. Moreover, the processing overhead in the FNs, small-cell BSs, APs, and user devices is much smaller than that in the mobility management entity (MME) and the core network. Thus, the fog-computing enabled handover procedure is able to achieve significant reductions in signaling overhead and in data traffic as compared with the conventional RAN and the centralized cloud RAN, and the long transmission delay and heavy burden on the fronthaul (usually seen in cloud RANs) can be effectively alleviated.

5.2.3 *Mobility Robustness and Handover Optimization*

Fog-computing enabled wireless networks are expected to build on SDN and NFV to reduce the management and configuration costs and to improve the scalability and resource utilization of the wireless network [7]. Each user device is associated with a clone in a FN or in the cloud server, where a VM executes the requested applications for the user device [19]. MRO becomes specially critical in fog computing, NFV, and SDN enabled RANs because the composition, decomposition, and migration processes of proxy VMs are determined by the locations of the proxy VMs' registered mobile devices [8]. When a mobile device moves from one BS's coverage area into another BS's coverage area, following the location updating procedure, it should report its current location to the MME in the OpenFlow control layer [8]. Thus, a proxy VM can always have the updated knowledge of the locations of its associated mobile devices, and will perform proxy VM decomposition, composition, and migration processes among the FNs accordingly.

Note that most of the memory in a proxy VM is used to store the semantic models and device profiles, which do not usually change after the initial installation. Some proxy VM migrations will not reduce the latency, but will increase the burden on the core network. Hence, it is not always necessary to migrate the proxy VM when a mobile device moves into the coverage area of a new BS. Meanwhile, it has been reported that around 10–30% of all human movements are due to social relationships, and 50–70% are attributed to periodic behaviors [20]. Most users mainly settle in a limited number of areas covered by a few BSs. Therefore, the dynamics of human mobility can be reasonably predicted using mathematical models. Distributed cooperative Q-learning has been used to adjust the handover settings in response to mobility pattern changes in wireless networks [21]. Based on such predictions, it is possible to pre-allocate replicas of users' proxy VMs in the optimal FNs. For example, the replicas of a mobile device's semantic models can be pre-allocated to the FNs that are connected to the BSs frequently visited by the user, e.g., the BS serving the user's home or workplace. In this case, if a proxy VM needs to migrate to a different FN that contains a replica of the proxy VM, then only the difference between the proxy VM migration and its replica needs to be transferred, instead of the whole memory of the proxy VM. Thus, the time duration and traffic load of the proxy VM migration can be significantly reduced.

5.3 Self-Coordination of Inter-Cell Interference

In a multi-tier HetNet, different tiers of BSs (or APs) can either operate in a dedicated frequency band each or share the same frequency band(s). In the former case, inter-cell interference would only occur within the same tier, i.e., there is no cross-tier interference; however, given the limited total available frequency spectrum for wireless communications, the frequency bandwidth (and thus the capacity) available to each tier of cells would be largely limited. In the latter case, when there are a large number of densely deployed small-cell BSs and APs, the cross-tier co-channel interference between cells will be the dominating performance-limiting factor to the multi-tier HetNet. Over the past few years, the co-channel deployment of multi-tier HetNets has received a higher popularity than the dedicated-channel deployment [18, 22, 23].

If inter-cell interference is not effectively mitigated, then it would be impossible to achieve the high capacity, high area spectral efficiency, high energy efficiency, or any other benefit promised by the dense deployment of HetNets [18]. Inter-cell interference coordination (ICIC) and enhanced ICIC (eICIC) are self-optimization functions that aim to minimize the mutual interference between neighboring cells sharing the same radio spectrum by adaptively coordinating the radio resource allocation and/or user scheduling among neighboring cells. In inter-cell interference limited wireless networks, ICIC should be performed in both the uplink and the downlink for the data channels.

Autonomous ICIC requires coordinated radio resource allocation among neighboring cells, relying on frequent signaling among the cells via the X2 interface. For instance, the signaling of the high interference indicator (HII) and the relative narrowband transmit power (RNTP) between cells for proactive ICIC, and the signaling of the overload indicator (OI) between cells for reactive ICIC [13].

In fog-computing enabled RANs, FNs are distributed and are connected to the BBU pool via fronthaul links. Cooperative radio signal processing and coordinated radio resource management can be executed in FNs and their associated BSs and APs, which integrate not only the front radio frequency, but also caching, local distributed radio signal processing, and radio resource management capabilities [24]. The caching capabilities in fog-computing enabled edge devices can be used to improve the spectral efficiency and the energy efficiency while maintaining a low latency level [25].

It is expected that fog-computing enabled RANs will work in a user-centric and/or content-centric manner, where mobile devices can be collectively served by all available nearby FNs and their associated BSs and APs. Fog computing will enable self-coordination among neighboring cells to avoid potential conflicts between multiple ICIC related SON functions that are executed at the same time, thus achieving user-centric and/or content-centric networking among collocated small cells and macrocells. In [26], a distributed coalitional game based cluster formation algorithm, where FNs and their associated BSs and APs can autonomously form clusters, was proposed to mitigate inter-cell interference and maximize the system throughput.

5.3.1 Interference-Aware Radio Resource Allocation

Game theory has been widely employed to alleviate co-channel inter-cell interference in RANs [26, 27]. Under the game-theoretic framework, a utility function can be defined according to the optimization objective and the characteristics of the system model considered, including the modeling of inter-cell interference. Accordingly, the optimization of radio resource allocation in a RAN can be formulated as the maximization of the defined utility function under all necessary constraints.

In a fog-enabled HetNet under the co-channel deployment of all tiers, the optimization of uplink subchannel and transmit power allocation can be modeled as a non-cooperative game to achieve interference-aware allocation of radio resources, considering the selfish and rational behavior of fog-enabled network nodes and user devices [11]. As each fog-enabled user device has some caching capabilities, a reward function can be defined to encourage the user devices to participate in network edge caching [28]. Meanwhile, to mitigate the uplink interference caused by user devices to their neighboring BSs, a convex pricing function that is proportional to the transmit power of each user device can be introduced into the user device's utility function. Accordingly, in the non-cooperative game, the net utility function of a fog-enabled user device is defined as its maximum achievable data rate added by the reward function for the caching offered by it and subtracted by the pricing function for the uplink interference caused by it. The computational complexity of the joint optimization problem can be lowered by decomposing it into two subproblems: subchannel allocation and power allocation. The subchannel allocation is first optimized by maximizing the net utility function for each user device. Given the optimized subchannel allocation, the power allocation problem can be modeled as a super-modular game [11], for which the existence of Nash equilibrium in each individual subchannel has been proven [29]. The Nash equilibrium can be reached by devising an iterative algorithm, where the transmit power of each user device is initialized at the smallest possible power level and is then iteratively updated following the super-modular game.

To further discuss fog-enabled interference-aware radio resource allocation, let us consider a simple scenario where the fog-enabled HetNet contains only one high-power macrocell BS and a number of low-power fog-enabled small-cell BSs uniformly distributed in the coverage area of the macrocell BS. Within the coverage area of each cell, the associated user devices are uniformly distributed. The small cells share the same frequency band with the macrocell. Each fog-enabled small-cell BS makes decisions on the subchannel allocation and the power allocation in each subchannel for the user devices associated with it. For each subchannel, the wireless channel model consists of distance-dependent path loss and Rayleigh flat fading.

Figure 5.3 (see Fig. 6 in [11]) shows the total net utility of all user devices obtained from simulations in the above considered fog-enabled HetNet (denoted by "FRAN") and that of a conventional HetNet (denoted by "non-FRAN") versus the number of user devices served per small cell, for different numbers of small cells per macrocell coverage area, where small cells are denoted by femtocells and F-APs in non-FRAN and FRAN, respectively, while user devices are denoted by femto users and F-UEs in non-FRAN and FRAN, respectively. It can be observed that the fog-enabled interference-aware resource allocation scheme (denoted by "proposed scheme for FRAN" in Fig. 5.3) significantly outperforms the conventional resource allocation in the HetNet. The total net utility of all user devices in the fog-enabled HetNet generally increases with the number of user devices per small cell, as well as increases with the number of small cells per macrocell coverage area. This is because based on the above described non-cooperative game between fog-enabled user devices, the interference pricing function in the net utility function of a user

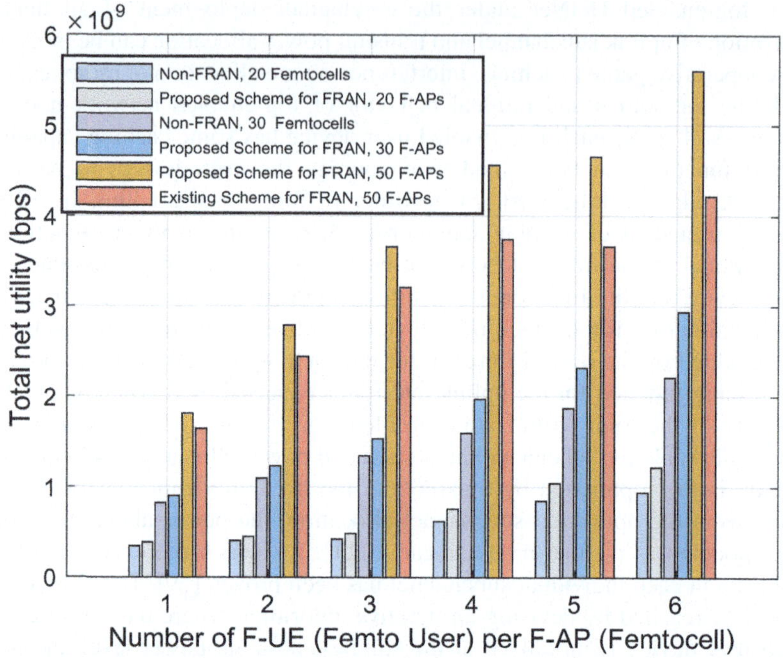

Fig. 5.3 Total net utility of all user devices in fog-enabled HetNets (denoted by "FRAN") and in conventional HetNets (denoted by "non-FRAN") versus the number of user devices per small cell [11]

device can effectively mitigate the uplink interference caused by user devices to the nearby BSs, while the caching reward function in the net utility function encourages fog-enabled user devices to offer local caching, thus relieving the traffic burden on both the fronthaul and the backhaul and improving the total capacity of the fog-enabled HetNet.

5.4 Self-Optimization of Coverage and Capacity

Coverage and capacity optimization (CCO) is a self-optimization function that aims to find the best trade-off between coverage and capacity, either network wide or per cell. The CCO associated optimization objectives include coverage, cell throughput, cell edge throughput, etc. To achieve these optimization objectives, CCO is usually combined with the optimization for ICIC and user scheduling. For example, coverage optimization may be devised through coverage mapping, coverage hole detection, or the detection of excessive interference spots.

Similar to autonomous ICIC, CCO functionalities can be executed in FNs and their associated BSs and APs, while the CCO objectives can be achieved by the dynamical arrival and departure (or switch on and off) of FNs and their associated BSs and APs [30]. The service orchestration for FNs, which covers automated instantiation, replication, and migration of service instances on all the associated FNs, can be logically centralized at the SDN controller [4]. In this case, the SDN controller needs to collect and maintain up to date information about all the FNs, network edge nodes, and end devices that are associated with it. The fog-node features that need to be continuously or periodically updated include their available memory resources, available storage, operating system(s), and software applications [5]. The capabilities and status of network edge nodes that need to be kept updated include the RATs used, link capacities, residual bandwidth, neighbor lists, etc. As for the connected end devices, the SDN controller needs to know their supported RATs, the types of services that they request, and the locations and moving speeds of mobile devices.

Moreover, CCO can exploit the diversity offered by the four possible transmission modes in a fog-computing enabled RAN: device-to-device (D2D) and relay mode, local distributed coordination mode, global centralized mode, and macrocell mode [11]. The CCO functions running in FNs or network edge nodes enable each associated user device to select the most appropriate transmission mode while considering the user device's movement speed, wireless communication range, geographic location, QoS and/or QoE requirements, computing and processing capabilities, and caching capability.

5.4.1 Deep-Learning Enabled Coverage and Capacity Optimization

As a key enabling technology for the fifth generation (5G) of wireless communication systems, massive MIMO using large numbers of transmitting and receiving antennas has been widely expected to enhance the system performance in terms of spectrum efficiency and energy efficiency [31–33]. In a wireless network equipped with massive MIMO, there will be a huge number of CCO related system parameters, including reference signal power levels, antenna tilting parameters, scheduling parameters, etc. These make it very difficult and expensive to automate CCO [34].

In the downlink of a multi-user massive MIMO network, the number of transmit antennas at a base station is much larger than the total number of receive antennas of all the scheduled user devices in the cell. In view of this, it would be more realistic to achieve CCO by properly tuning the user scheduling parameters than optimizing the antenna tilting parameters [34]. It is worth noting that the spectrum efficiency and capacity of a multi-user MIMO network is mainly determined by a small number of users with low SINR values, e.g., the cell-edge users. While scheduling the cell-center users and the cell-edge users simultaneously, in order to avoid the network capacity being lowered by the cell-edge users, it is necessary to align the users' SINR values in the scheduled user group. To this end, deep reinforcement learning based user scheduling schemes have been proposed to realize CCO [34].

Reinforcement learning provides a model-free method to solve a Markov decision process (MDP). An MDP typically consists of a state space S, an action space A, a stationary transition distribution describing the environment dynamics $p(s_{t+1}|s_t, a_t)$, which satisfies the Markov property, and a reward function r_t, where t denotes the timestep. In a general reinforcement learning setting, an agent observes a state s_t of the environment and chooses an action a_t at timestep t. As a result of the action, the state of the environment transits to s_{t+1} and the agent receives a reward r_t. Although the agent has no a priori knowledge of the environment, it can learn to take actions to maximize the expected cumulative discounted reward by randomly choosing actions and observing the transitions of the environments.

At each timestep t, an agent's decision-making procedure is characterized by a policy, $\pi(s, a) = Pr\{a_t = a|s_t = s\}, \forall s \in S, a \in A$, which is the probability that the agent takes action a in state s. In practical problems, the state space and action space are usually very large, making it intractable to store the policy in a tabular form. Consequently, function approximators have been used to parameterize the policy as $\pi_\theta(s, a)$ with a parameter vector θ, where $\theta \in \Re^l$, for $l << |S|$. Then, training of the policy can be performed by following the gradient of the cumulative discounted reward with respect to the parameter vector [35].

In the CCO orientated user scheduling scheme proposed in [34], in each transmission time interval, the minimum required user SINR threshold $SINR_{\min}$ and the users' aligned signal strength R are dynamically configured by a deep reinforcement learning algorithm. More specifically, each sector of the cell is considered as an agent that aims to maximize both the cell average spectrum efficiency and the cell-edge spectrum efficiency. The action space is constructed by the combinations of all possible discrete levels of $SINR_{\min}$ and R. A deep neural network (DNN) is used as the function approximator to compute the policy that the agent should follow in a state, where the policy is chosen as the combination of $SINR_{\min}$ and R with the largest probability of leading to the largest reward. The advantage of using a DNN is that it does not require human crafted features. Extensive simulation results have shown that the deep reinforcement learning based user scheduling scheme can successfully track the dynamics in a multi-user massive MIMO network, thus effectively achieving CCO [34].

5.5 Self-Optimization of Network Resources

Despite the continuing proliferation and developments of smart mobile devices, they are still largely battery power limited and resource constrained, especially when the number of computationally intensive mobile applications are increasing at an even faster rate. These mobile applications would consume a large amount of energy and computing capacity, and impose stringent constraints on processing and response delays. To deliver satisfactory user experience, it has become necessary to offload the computationally intensive tasks from mobile devices to more powerful nodes of the mobile network infrastructure.

Future user-centric wireless networks should be able to autonomously optimize the resource provisioning for individual mobile applications at their runtime. In current 4G wireless networks, it is difficult to simultaneously achieve user-centric resource provisioning and network-wide load balancing, due to the lack of flexibility in network operation [36]. To address this issue, the emerging NFV technology slices a wireless network into dedicated end-to-end (E2E) virtual networks, each containing a chain of virtual network functions (VNFs) to deliver a customized or personalized service. VNFs implement network functions (such as load balancing, routing, caching, security, etc.) using software and are connected through service function chaining to create and deliver communication services within a NFV infrastructure, which consists of both virtual and physical resources.

In fog-computing enabled RANs, resources are not just limited to radio resources, but also include the caching and computing capabilities in network edge devices. Accordingly, the resource management in fog-computing enabled RANs goes beyond the traditional radio resource allocation to include also the allocation of caching and computing resources at the network edge. As a result, future fog-computing enabled HetNets would be different in networking, computing, storage, and control as compared with conventional RANs and the cloud RAN. The seamless integration of multiple RATs in HetNets would rely on the virtualization of computational, storage, and networking resources [37]. For the optimization of resource allocation and utilization, the objective would typically be to maximize the overall energy efficiency for computing and networking, under the system constraints and QoS or QoE requirements imposed by users and things [11]. In [38], the weighted sum performance improvement in time saving and energy reduction of the system was maximized by jointly optimizing the offloading decisions, the allocation of local computational resources, and wireless transmission power.

5.5.1 Network Edge Caching

It has been observed that with the increasing popularity of location-based mobile applications, a lot of social application data exchanged between neighboring user devices shows a high degree of conformity. Some social applications would only generate data traffic between user devices in physical proximity. Besides, users from the same social group or having the same social interest(s) may request the same contents over the downlink during a certain period of time.

In fog-computing enabled wireless networks, FNs integrate not only the fronthaul radio frequency but also the physical-layer signal processing functionalities and relevant procedures of the upper layers. Thus, FNs can implement collaborative radio signal processing locally using their adequate computing capabilities and can manage their storage and caching memories flexibly. Thus, a FN is able to collect (or offload), store, process, and manage the application data and tasks from the network edge nodes and mobile devices that are connected to it. For location-based and/or social mobile applications, FNs and their associated network edge nodes can proactively prefetch and cache the contents that are highly locally popular and/or relevant, so as to maximize the overall energy efficiency in mobile service provisioning while guaranteeing the QoS and QoE requirements of users and things.

5.5.2 Computation Offloading

Computation offloading is a procedure that migrates resource-intensive computation task(s) from a user device to the nearby resource-rich network node(s) [39]. Computation offloading decisions, such as whether or not to offload a computation task, whether to fully or partially offload a computation task, and how to perform the offloading, need to be optimized. Depending on the specific mobile application that a computation task belongs to, there are three main criteria for computation offloading decision-making [40]: (1) whether the application contains certain parts that cannot be offloaded (e.g., user input, camera functions, or position acquisition that need to be executed at the user device); (2) whether or not the application requires continuous execution, which makes it difficult to estimate the amount of data to be processed; and (3) whether or not there exists a mutual dependency between individual parts to be processed.

Most existing computation offloading decision algorithms aim to minimize the energy consumption at mobile devices while meeting the maximum acceptable execution delay of the offloaded application, or to find an optimal trade-off between the energy consumption and the execution delay for a mobile application [40]. The allocation of computational and networking resources was jointly optimized to minimize the energy consumption while satisfying the latency requirements in [41]. In [42], the energy consumption of a user device was minimized by jointly optimizing the allocation of computational and radio resources in the offloading process, where the processing of an application is partitioned between the user device and a nearby RAN node. The total energy consumption of the user device includes the energy consumed in the uplink transmission and downlink reception, as well as the energy consumed for local processing at the user device. Meanwhile, the execution of the application must finish within a time limit corresponding to a given QoS requirement. The trade-offs between the energy consumption and the transmission and processing delays for local processing and for computation offloading (i.e., remote processing) were analyzed in [43, 44].

Fog computing, in conjunction with mobile cloud computing [45], has been considered by both the industry and the academia as a promising way to enable energy-efficient and low-latency computation offloading for mobile devices. Fog-computing enabled wireless networks can readily support and further improve the performance of computation offloading by properly forming a cluster of neighboring FNs (or a local fog network) and optimally distributing the computation tasks among the selected neighboring FNs and/or the cloud server, i.e., to optimize the task distribution across the fog and the cloud [30]. The scheduling of computation tasks should be optimized adaptively according to whether all necessary information about the formed local fog network is completely known to all the involved FNs; or for a certain computation task, only partial (or even none) knowledge of the computing and processing capabilities, availability of computing and networking resources, or service loads of some involved FNs is available to the other FNs [30]. The computation offloading decisions and the associated allocation of computing, storage, and networking resources are optimized in the FNs to achieve performance improvements for the wireless network [41] and/or for individual user devices, e.g., improved energy efficiency and reduced delay.

Computation offloading from a mobile device to the cloud or to the fog can be performed on the coarse-grained application level [41, 46], the fine-grained task level [38, 47, 48], or a combination of the above two levels [42, 49]. The computation offloading decisions can be made in a centralized manner or in a distributed manner [50, 51], for individual user devices separately or jointly for multiple user devices all together [38]. If computation offloading is performed for a single user device, then the computation tasks need to be partitioned and an offloading decision needs to be made for each task (i.e., whether or not to offload it) [43, 47, 48]. Task partitioning and assignment have been studied for data partitioned oriented applications [42], of which the amount of data to be processed is known a priori and different portions of the data can be processed in parallel. If computation offloading is performed jointly for multiple user devices at the same time, then the computational resources available at a FN (or a cloud server) and the networking resources between the FN (or the cloud server) and the multiple user devices need to be properly shared among the user devices. To optimize the computation offloading decisions in a multi-user multi-fog-node scenario, game theory based approaches can be used [50, 51].

5.5.3 *Joint Optimization of Computation Offloading and Resource Allocation*

In fog-computing enabled wireless networks, the computation offloading decisions can be jointly optimized with the allocation of computing resources, radio spectrum resources, and transmit power [52–54]. For future user-centric wireless networks, the design of a computation offloading scheme cannot just focus on system-level performance improvement, because the optimization of network-wide performance may not be able to guarantee the fairness among individual user devices. It is likely that only those user devices with good channel conditions (e.g., high channel gains, low interference levels, or both) can benefit from computation offloading, while user devices under bad channel conditions may even experience degraded QoS or QoE. It is thus critical to consider the QoS and/or QoE for each individual user device in the optimization of offloading decisions and the associated resource allocation.

For ease of discussion, let us consider a simple fog-computing enabled RAN scenario, which consists of $N (N \geqslant 1)$ user devices, one FN, and one cloud server. Each user device is connected to the FN via a wireless link. The FN is connected to the cloud server via a high-speed wired link. Each user device has one application to be either executed locally (i.e., by itself) or offloaded for remote processing in the FN or the cloud server. A computation offloading event is triggered by a user device transmitting an offloading request to the FN [49]. The request should include necessary information about the user device (e.g., its local processing capability and power level) and properties of the application (e.g., the maximum tolerable delay) [38]. According to all the received offloading requests from user devices and the instantaneous channel state information, the FN decides where each application should be processed, i.e., in the user device locally, in the fog, or in the cloud, and sends the offloading decision to the corresponding user device. For analytical tractability, it can be assumed that the delays due to offloading request queuing and decision-making are negligible, i.e., the FN decides the offloading strategy for all the received user requests at the beginning of an offloading period [55].

For many mobile applications, such as natural language processing and face recognition, where the input data size is relatively small so that the application offloading could be completed during a time shorter than the timescales of user device mobility and the dynamics of wireless networks [52]. The system can thus be assumed to be quasi-static, where all user devices and the wireless network remain stationary during an offloading period [50].

The set of all user devices is denoted by $\mathcal{N} = 1, \ldots, N$. If an application is offloaded from a user device to the FN for processing, then the user device is referred to as a fog-processing device. The set of all fog-processing user devices is denoted by $\mathcal{N}_1$. The total number of fog-processing user devices is given by $N_1 = |\mathcal{N}_1|$. If an application is offloaded from a user device to the cloud server for processing, then the user device is referred to as a cloud-processing device. All the fog-processing and cloud-executing user devices are collectively referred to as remote-processing user devices and are put in the set $\mathcal{N}_2$. The total number of remote-processing user devices is given by $N_2 = |\mathcal{N}_2|$.

For remote processing in the cloud server, an application needs to be first transmitted from a user device to the FN through a wireless link, and then forwarded by the FN to the cloud server through a wired link. Since the cloud server should have plenty of computing resources and the wired link between the FN and the cloud server typically has a sufficiently large capacity, the allocation of these resources among the remote-processing applications is less of a concern than the limited wireless communication resources, which need to be shared among all the remote-processing user devices for communicating with the FN. Denote the total radio frequency bandwidth by B Hz and the portion of bandwidth assigned to the nth remote-processing user device by a_n, where $n \in \mathcal{N}$, and $a_n \in [0, 1]$. To avoid interference between simultaneous transmissions to the FN, at any offloading period, orthogonal radio channels are allocated to different remote-processing user devices, i.e., $\sum_{n \in \mathcal{N}_2} a_n \leqslant 1$ [56, 57]. Since the output data after an application has been processed is typically of a small size, the optimization of offloading decisions and the associated resource allocation usually focuses on the uplink transmissions only [19, 38, 43, 50].

The application of the nth user device can be described by $J_n = \{D_n, App_n, \tau_n^{\max}\}$, $n \in \mathcal{N}$, where D_n denotes the size of input data (in bits), App_n denotes the required processing density (in CPU cycles/bit), which depends on the computational complexity of the application [49, 58], and $\tau_n^{\max}$ stands for the maximum tolerable latency (in seconds) [52]. The number of CPU cycles required to process the entire application J_n is given by $C_n = D_n App_n$ [42, 49]. In the following, we assume that the FN knows the values of D_n, C_n, and App_n, e.g., by employing program profilers [38, 59]. Supported by NFV and SDN technologies, the FN constructs a clone for each user device, i.e., the program for processing application J_n is backed up in the FN [43, 58] and can be downloaded by the cloud server through a high-speed wired link if needed [41]. Therefore, in case of cloud processing, only the data of D_n bits need to be transmitted from the nth user device to the cloud server. It should be noted that D_n, C_n, App_n, and $\tau_n^{\max}$ are inherent parameters of application J_n, and they do not change with where J_n is processed [52].

In the following, we will look into the delay and power consumption for local processing, fog processing, and cloud processing, respectively.

Local Processing

Let f_n^{loc} and p_n^{loc} ($n \in \mathcal{N}$) denote the local computation capability (in CPU cycles/s) and the local executing power consumption (in watts) of the nth user device, respectively. The delay and energy consumption for locally processing application J_n at the nth user device are given, respectively, by [52]

$$T_n^{loc} = C_n / f_n^{loc}, \tag{5.1}$$

$$E_n^{loc} = p_n^{loc} (C_n / f_n^{loc}). \tag{5.2}$$

Fog Processing

If application J_n is to be processed in the FN, then the nth user device needs to transmit the input data of size D_n to the FN through a wireless link, and the FN needs to allocate sufficient computing resources (in CPU cycles/s) for processing application J_n. The achievable transmission rate of the nth user device is given by [52]

$$r_n = a_n B \log_2 \left(1 + \frac{p_n^{loc}}{a_n N_0 B}\right), \tag{5.3}$$

where p_n^{com} is the transmission power of the nth user device, h_n is the channel power gain between the nth user device and the FN, and N_0 is the additive white Gaussian noise (AWGN) power spectral density (in mW/Hz).

Under the assumption that the fog processing for an application starts only after all the input data has been received by the FN, the delay and energy consumption of fog processing for application J_n are given, respectively, by [52]

$$T_n^{fog} = D_n/r_n + C_n/f_n^{fog}, \tag{5.4}$$

$$E_n^{fog} = p_n^{com}(D_n/r_n) + p_n^{id}(C_n/f_n^{fog}), \tag{5.5}$$

where f_n^{fog} (in CPU cycles/s) denotes the computing resources allocated by the FN to application J_n, and p_n^{id} denotes the power consumption (in watts) of the nth user device in the idle mode.

Cloud Processing

If application J_n is to be offloaded to the cloud server, then the nth user device needs to first transmit the input data of size D_n through a wireless link to the FN, which then forwards the received input data to the cloud server through a wired link. Denote the data rate of the wired link allocated to the nth user device by R_n^{fc} (in bits/s), and the cloud processing capability assigned to application J_n by f_n^c (in CPU cycles/s). Then, the delays caused by the wired transmission and cloud processing are given by $T_n^{fc} = D_n/R_n^{fc}$ and $T_n^c = C_n/f_n^c$, respectively. Accordingly, the total delay and total energy consumption of cloud processing for the nth user device are given, respectively, by [52]

$$T_n^{cloud} = D_n/r_n + T_n^{fc} + T_n^c, \tag{5.6}$$

$$E_n^{cloud} = p_n^{com}(D_n/r_n) + p_n^{id}(T_n^{fc} + T_n^c). \tag{5.7}$$

The computation offloading decision for the nth user device can be represented by the following three binary indicators, $x_n, y_n, z_n \in \{0, 1\}$, where $x_n = 1$, $y_n = 1$, and $z_n = 1$ indicate that application J_n is processed by the nth user device itself locally, by the FN, and by the cloud server, respectively; otherwise, $x_n = 0$, $y_n = 0$, and $z_n = 0$. The three binary indicators are subject to the following constraint:

$$x_n + y_n + z_n = 1, \forall n \in \mathcal{N}, \tag{5.8}$$

which implies that application J_n is processed at one and only one location selected from the nth user device, the FN, and the cloud server.

Therefore, the delay and energy consumption for processing application J_n in the above considered fog-computing enabled RAN scenario can be expressed, respectively, as [52]

$$T_n = T_n^{loc} x_n + T_n^{fog} y_n + T_n^{cloud} z_n, \tag{5.9}$$

$$E_n = E_n^{loc} x_n + E_n^{fog} y_n + E_n^{cloud} z_n. \tag{5.10}$$

Accordingly, the cost of the nth user device can be defined as the weighted sum of delay and energy consumption as

$$Cost_n = \lambda_n^t T_n + \lambda_n^e E_n, \tag{5.11}$$

where $\lambda_n^e, \lambda_n^t \in [0, 1], n \in \mathcal{N}$, denote the weights of energy consumption and delay for the nth user device, respectively.

One way to ensure the fairness among all user devices is to minimize the maximum cost among all the user devices, without exceeding the maximum tolerable delay for each mobile application. This involves the joint optimization of the computation offloading decisions for all user devices, $\boldsymbol{\pi} = [\mathbf{x}, \mathbf{y}, \mathbf{z}] = [x_1, \ldots, x_N, y_1, \ldots, y_N, z_1, \ldots, z_N]$, the fog-node computing resource allocation among fog-processing user devices, $f^{fog} = [f_1^{fog} \ldots, f_N^{fog}]$, the assignment of wireless channel bandwidth to remote-processing user devices, $\mathbf{a} = [a_1, \ldots, a_N]$, and the transmission power allocation to remote-processing user devices, $\mathbf{p}^c om = [p_1^{com}, \ldots, p_N^{com}]$.

The above described fairness-aware maximum-cost minimization problem is formulated as [52]

$$(\mathscr{P}_1): \quad \min_{\boldsymbol{\pi}, \mathbf{f}^{fog}, \mathbf{p}^{com}, a} \quad \max_{n \in \mathcal{N}} Cost_n \tag{5.12}$$

Subject to

$$\begin{aligned}
&(C1): x_n, y_n, z_n \in \{0, 1\}, \forall n \in \mathcal{N}, \\
&(C2): x_n + y_n + z_n = 1, \forall n \in \mathcal{N}, \\
&(C3): \quad \textstyle\sum_n \in \mathcal{N} f_n^{fog} \leqslant F^{fog}, \\
&(C4): \quad f_n^{fog} \geqslant 0, \forall n \in \mathcal{N}, \\
&(C5): \quad 0 < a_n \leqslant 1, \forall n \in \mathcal{N}, \\
&(C6): \quad \textstyle\sum_n \in \mathcal{N}_2 a_n \leqslant 1, \\
&(C7): 0 \leqslant p_n^{com} \leqslant p_n^{\max}, \forall n \in \mathcal{N}, \\
&(C8): \quad T_n \leqslant \tau_n^{\max}, \forall n \in \mathcal{N},
\end{aligned}$$

where F^{fog} is the total computation capacity of the FN, $p_n^{\max}$ is the maximum allowed transmission power of the nth user device; if $x_n = 1$, i.e., local processing, then $p_n^{com} = 0$ and $a_n = 0$; if $x_n = 1$ or $z_n = 1$, i.e., non-fog processing, then $f_n^{fog} = 0$; (C1) and (C2) are the constraints on the binary offloading indicators for each user device; (C3) indicates that the allocated fog computing resources cannot exceed the total computation capability of the FN; (C4) is the non-negative constraint on fog computing resource allocation; (C5) and (C6) are the constraints on the wireless channel bandwidth allocation to remote-processing user devices; (C7) is the transmission power constraint of each user device; and (C8) is to guarantee that each application is executed within the maximum tolerable delay.

Note that the optimization problem ($\mathscr{P}_1$) is not convex due to the min–max formulation and the binary variables in $\boldsymbol{\pi}$ [52]. It is a mixed-integer non-linear programming problem, which is NP hard in general [60]. To reduce the computational complexity, the optimization problem ($\mathscr{P}_1$) can be first transformed into a quadratically constrained quadratic programming (QCQP) problem, which can then be converted into a standard convex problem via semidefinite relaxation [52]. The resulting problem can be solved in polynomial time using a standard convex optimization tool such as SeDuMi [61]. The above transformation, conversion, and solution are summarized in the computation offloading and resource allocation (CORA) algorithm [52, Algorithm 1].

In [52], the performance of the CORA algorithm was evaluated through Monte Carlo simulation in comparison with the following three benchmarking algorithms: (1) The offloading-only algorithm [50], where only offloading decisions are optimized to minimize the weighted sum of energy consumption and delay for each user device, without optimizing the resource allocation. (2) The resource-only algorithm [56], where only the allocation of resources (including transmission power, frequency bandwidth, and computing resources) is optimized to minimize the power consumption of each user device, without optimizing offloading decisions. (3) Local-only processing, where all user devices process their applications themselves without any optimization for offloading decisions or resource allocation.

The scenario for simulation consists of one FN, one cloud server, and N user device, as well as the TGn path loss model [62] and Rician fading with a 6 dB Rician factor [63]. Unless otherwise mentioned, the simulation parameters are set as follows [52]: $N = 6$, $B = 15\,\text{MHz}$, $N_0 = -174\,\text{dBm/Hz}$, $F^{fog} = 2\text{G cycles/s}$, $p_n^{\max} = 0.1\,\text{W}$, p_n^{id} and p_n^{loc} are independently and uniformly distributed in the range of 0.001–0.01 W and in the range of 0.1–0.5 W, respectively, $f_n^c = 4\text{G cycles/s}$, f_n^{loc} is uniformly distributed in the range of 0.5–1.5G cycles/s, $\tau_n^{\max} = 4\,\text{s}$, $D_n = 0.42\,\text{MB}$, $App_n = 297.62\,\text{cycles/bit}$, and $R_n^{fc} = 1\text{M bits/s}$, $\forall n \in \mathscr{N}$.

Figure 5.4 (see Fig. 7 in [52]) plots the maximum energy consumption among all user devices versus the number of user devices for the four algorithms under comparison. In this simulation, the optimization objective of the CORA algorithm was set to minimize the energy consumption by setting $\lambda_n^e = 1$ and $\lambda_n^t = 0$, $n \in \mathscr{N}$. We can see that the maximum user-device energy consumption increases with the number of user devices for all the four algorithms, with the CORA algorithm achieving the lowest rate of increase. For any given number of user devices, the CORA algorithm always achieves the lowest maximum user-device energy consumption among the four algorithms.

Figure 5.5 (see Fig. 9 in [52]) shows the maximum energy consumption and the maximum delay among all user devices of the four algorithms under comparison for four different applications, which are the m-queens puzzle for $m = 4, 5, 6$, and 7 [46, 58]. The four applications possess the same size of data, i.e., $D_n = 200\,\text{KB}$ for each given m, but with different sizes of processing density, where $App_n = 87.8$, 263, 1760, and 8250, for $m = 4, 5, 6$, and 7, respectively. We can see that both of the maximum user-device energy consumption and delay increase with m, with the CORA algorithm achieving the lowest maximum user-device energy consumption and the shortest maximum user-device delay for each given m.

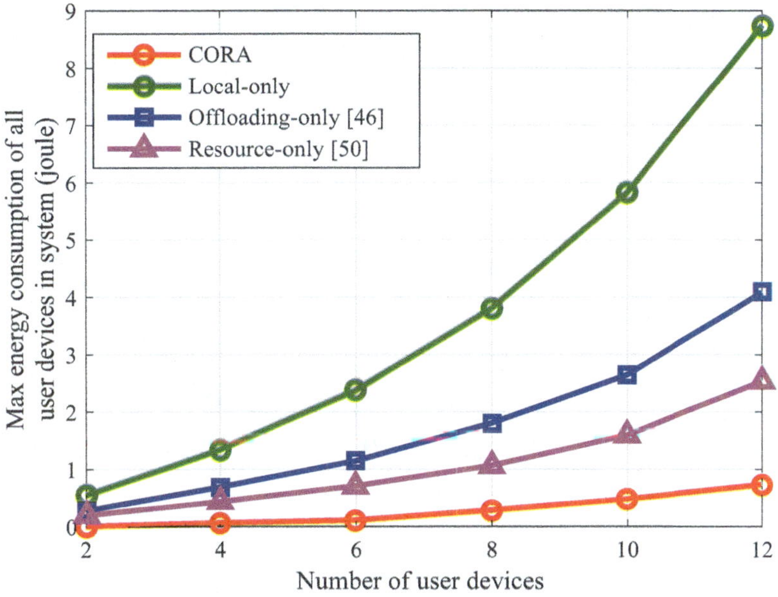

Fig. 5.4 The maximum energy consumption among all user devices versus the number of user devices [52]

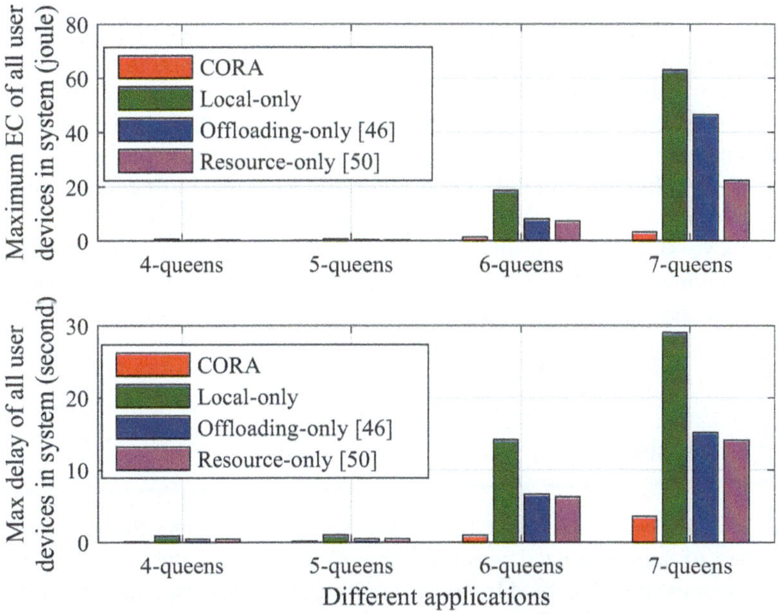

Fig. 5.5 The maximum energy consumption and maximum delay among all user devices for four different applications [52]

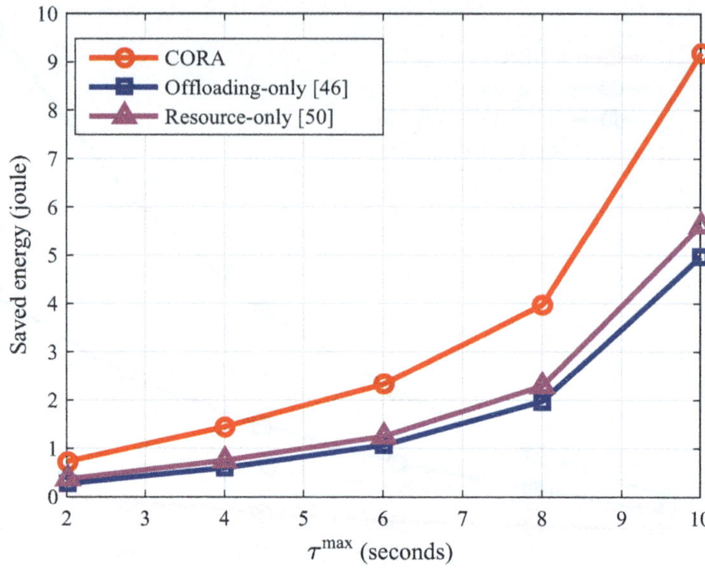

Fig. 5.6 The maximum saved energy among all user devices with respect to the local-only processing versus the maximum tolerable delay $\tau_n^{\max}$ (s) [52]

Figure 5.6 (see Fig. 11 in [52]) shows the energy saving achieved by the CORA algorithm, the offloading-only algorithm [50], and the resource-only algorithm [56] with respect to the local-only case versus the maximum tolerable delay $\tau_n^{\max}$. We can see that for all the three algorithms shown, the energy saving increases with the maximum tolerable delay. This is because a looser delay constraint will lead to more offloaded mobile applications, and thus more conserved energy for the user devices. For any given value of $\tau_n^{\max}$, the CORA algorithm saves the most energy among the three algorithms.

The above simulation results show that the joint optimization of offloading decision-making and resource allocation in the CORA algorithm is able to effectively reduce the energy consumption and delay for user devices in a fog-computing enabled RAN.

5.6 Conclusion

In this chapter, we have looked into the various new opportunities and new capabilities of self-optimized wireless networking, e.g., scalable and on-demand mobility management, enabled by fog computing. Recent works have demonstrated that fog computing, in conjunction with the emerging software-defined networking and network function virtualization technologies, allows the self-optimized dynamic relocation of computing, caching, and networking resources across the cloud, the fog, the network edge, and the things, as well as self-optimized management of network functions and mobile applications. Fog-computing enabled wireless networks provide user devices with distributed local caching, computation offloading, collaborative radio signal processing, and cooperative resource management at the edge of the network.

References

1. Moysen J, Giupponi L (2017) From 4G to 5G: self-organized network management meets machine learning, arXiv:1707.09300v1 [cs.NI]
2. 3GPP (2016) Technical specification group services and system aspects; telecommunication management; self-organizing networks (SON); concepts and requirements (Release 13). Technical Report TS 32.500, v13.0.0
3. Bonomi F, Milito R, Zhu J, Addepalli S (2012) Fog computing and its role in the Internet of things. In: Proceedings of the first edition of the MCC workshop on mobile cloud computing, Helsinki, August 2012, pp 13–16
4. Tomovic S, Yoshigoe K, Maljevic I, Radusinovic I (2017) Software-defined fog network architecture for IoT. Wirel Pers Commun 92(1):181–196
5. Bonomi F, Milito R, Natarajan P, Zhu J (2014) Fog computing: a platform for Internet of things and analytics. In: Big data and internet of things: a roadmap for smart environments, vol 546. Springer International Publishing, Cham, pp 169–186
6. Chih-Lin I, Yuan Y, Huang J, Ma S, Cui C, Duan R (2015) Rethink fronthaul for soft RAN. IEEE Commun Mag 53(9):82–88
7. Liang K, Zhao L, Chu X, Chen H (2017) An integrated architecture for software defined and virtualized radio access networks with fog computing. IEEE Netw 31(1):80–87
8. Sun X, Ansari N (2016) EdgeIoT: mobile edge computing for the Internet of things. IEEE Commun Mag 54(12):22–29
9. Yazici V, Kozat UC, Sunay MO (2014) A new control plane for 5G network architecture with a case study on unified handoff, mobility, and routing management. IEEE Commun Mag 52(11):76–85
10. Zhang H, Long K, Chu X, Aghvami H, Leung V (2017) Network slicing based 5G and future mobile networks: mobility, resource management, and challenges. IEEE Commun Mag 55(8):138–145
11. Zhang H, Qiu Y, Chu X, Long K, Leung V (2017) Fog radio access networks: mobility management, interference mitigation and resource optimization. IEEE Wirel Commun 24(6):120–127
12. McKeown N, et al (2008) OpenFlow: enabling innovation in campus networks. ACM SIGCOMM Comput Commun Rev 38(2):69–74
13. 3GPP (2013) Technical specification group radio access network; evolved universal terrestrial radio access network (EUTRAN); X2 application protocol (X2AP) (Release 11). Technical Report TS 36.423, v.10.7.0
14. 3GPP (2014) Study on enhancements of OAM aspects of distributed self-organizing networks (SON) functions (Release 12). Technical Report TR 32.860
15. Mwanje S, Mitschele-Thiel A (2013) Minimizing handover performance degradation due to LTE self organized mobility load balancing. In: IEEE VTC Spring, Dresden, 2–5 June 2013, pp 1–5
16. Lopez-Perez D, Guvenc I, Chu X (2012) Mobility management challenges in 3GPP heterogeneous networks. IEEE Commun Mag 50(12):70–78
17. Lopez-Perez D, Guvenc I, Chu X (2012) Theoretical analysis of handover failure and ping-pong rates for heterogeneous networks. In: IEEE ICC'12 WS, Ottawa, 10–15 June 2012, pp 6774–6779

18. Chu X, Lopez-Perez D, Yang Y, Gunnarsson F (eds.) (2013) Heterogeneous cellular networks – theory, simulation and deployment. Cambridge University Press, Cambridge, pp. 1–494. ISBN-13: 9781107023093
19. Mao Y, Zhang J, Letaief KB (2016) Dynamic computation offloading for mobile-edge computing with energy harvesting devices. IEEE J Sel Areas Commun 34(12):3590–3605
20. Cho E, Myers S, Leskovec J (2011) Friendship and mobility: user movement in location-based social networks. In: Proceedings of the 17th ACM SIGKDD, San Diego, CA, August 2011, pp 1082–1090
21. Mwanje S, Mitschele-Thiel A (2014) Distributed cooperative Q-learning for mobility-sensitive handover optimization in LTE SON. In: IEEE ISCC
22. Chu X, Wu Y, Lopez-Perez D, Tao X (2011) On providing downlink services in collocated spectrum-sharing macro and femto networks. IEEE Trans Wirel Commun 10(12):4306–4315
23. Zhang H, Jiang C, Beaulieu NC, Chu X, Wen X, Tao M (2014) Resource allocation in spectrum-sharing OFDMA femtocells with heterogeneous services. IEEE Trans Commun 62(7):2366–2377
24. Ai Y, Peng M, Zhang K (2017) Edge cloud computing technologies for Internet of things: a primer. Digital Commun Netw. https://doi.org/10.1016/j.dcan.2017.07.001
25. Jia S, Ai Y, Zhao Z, Peng M, Hu C (2016) Hierarchical content caching in fog radio access networks: ergodic rate and transmit latency. China Commun 13:1–14
26. Sun Y, Dang T, Zhou J (2016) User scheduling and cluster formation in fog computing based radio access networks. In: IEEE ICUWB, Nanjing, 16–19 October 2016
27. Zhang H, Jiang C, Beaulieu N, Chu X, Wang X, Quek TQS (2015) Resource allocation for cognitive small cell networks: a cooperative bargaining game theoretic approach. IEEE Trans Wirel Commun 14(6):3481–3493
28. Liang C, Yu FR, Yao H, Han Z (2016) Virtual resource allocation in information-centric wireless networks with virtualization. IEEE Trans Veh Technol 65(12):9902–9914
29. Zhang H, Chu X, Zheng W, Wen X (2012) Interference-aware resource allocation in co-channel deployment of OFDMA femtocells. In: IEEE ICC'12, Ottawa, 10–15 June 2012, pp 4663–4667
30. Lee G, Saad W, Bennis M (2017) An online optimization framework for distributed fog network formation with minimal latency, arXiv:1710.05239v1 [cs.IT]
31. Ge X, Huang M, Chen J, Xu H, Xu J, Zhang W, Yang Y (2016) Wireless single cellular coverage boundary models. IEEE Access 4:3569–3577
32. Ge X, Qiu Y, Chen J, Huang M, Xu H, Xu J, Zhang W, Yang Y, Wang C, Thompson J (2016) Wireless fractal cellular networks. IEEE Wirel Commun 23(5):110–119
33. Ge X, Zi R, Xiong X, Li Q, Wang L (2017) Millimeter wave communications with OAM-SM scheme for future mobile networks. IEEE J Sel Areas Commun 35(9):2163–2177
34. Yang Y, Li Y, Li K, Zhao S, Chen R, Wang J, Ci S (2018) DECCO: Deep-learning enabled coverage and capacity optimization for massive MIMO systems. IEEE Access 6:23361–23371
35. Sutton R, McAllester D, Singh S, Mansour Y (1999) Policy gradient methods for reinforcement learning with function approximation. In: Proceedings of the 12th international conference on neural information processing systems (NIPS), pp 1057–1063
36. Piamrat K, et al (2011) Radio resource management in emerging heterogeneous wireless networks. Comput Commun 34(9):1066–1076
37. Silva ID, et al (2015) Tight integration of new 5G air interface and LTE to fulfil 5G requirements. In: IEEE VTC-Spring
38. Lyu X, Tian H, Sengul C, Zhang P (2017) Multiuser joint task offloading and resource optimization in proximate clouds. IEEE Trans Veh Technol 66(4):3435–3447
39. Barbarossa S, Sardellitti S, Di Lorenzo P (2014) Communicating while computing: distributed mobile cloud computing over 5G heterogeneous networks. IEEE Signal Process Mag 31:45–55
40. Mach P, Becvar Z (2017) Mobile edge computing: a survey on architecture and computation offloading. IEEE Commun Surv Tutorials 19(3):1628–1656
41. Sardellitti S, Scutari G, Barbarossa S (2015) Joint optimization of radio and computational resources for multicell mobile-edge computing. IEEE Trans Signal Inf Process Netw 1(2): 89–103

42. Munoz O, Pascual-Iserte A, Vidal J (2015) Optimization of radio and computational resources for energy efficiency in latency-constrained application offloading. IEEE Trans Veh Technol 64(10):4738–4755
43. Zhang W, Wen Y, Wu D (2015) Collaborative task execution in mobile cloud computing under a stochastic wireless channel. IEEE Trans Wirel Commun 14(1):81–93
44. Barbera MV, Kosta S, Mei A, Stefa J (2013) To offload or not to offload? The bandwidth and energy costs of mobile cloud computing. In: IEEE INFOCOM, pp 1285–1293
45. Shi W, Cao J, Zhang Q, Li Y, Xu L (2016) Edge computing: vision and challenges. IEEE Internet Things J 3(5):637–646
46. Liu K, Zhang X, Huang Z (2016) A combinatorial optimization for energy-efficient mobile cloud offloading over cellular networks. In: IEEE GLOBECOM'16, pp 1–6
47. Kao Y-H, Krishnamachari B, Ra M-R, Bai F (2017) Hermes: latency optimal task assignment for resource-constrained mobile computing. IEEE Trans Mob Comput 16(11):3056–3069
48. Dinh TQ, Tang J, La QD, Quek TQS (2017) Offloading in mobile edge computing: task allocation and computational frequency scaling. IEEE Trans Commun 65(8):3571–3584
49. Wang Y, Sheng M, Wang X, Wang L, Li J (2016) Mobile-edge computing: partial computation offloading using dynamic voltage scaling. IEEE Trans Commun 64(10):4268–4282
50. Chen X (2015) Decentralized computation offloading game for mobile cloud computing. IEEE Trans Parallel Distrib Syst 26(4):974–983
51. Cardellini V, et al (2016) A game-theoretic approach to computation offloading in mobile cloud computing. Math Program 157(2):421–449
52. Du J, Zhao L, Feng J, Chu X (2018) Computation offloading and resource allocation in mixed fog/cloud computing systems with min-max fairness guarantee. IEEE Trans Commun 66(4):1594–1608
53. Du J, Zhao L, Chu X, Yu F, Feng J, Chih-Lin I (2019) Enabling low-latency applications in LTE-A based mixed fog/cloud computing systems. IEEE Trans Veh Technol 68(2):1757–1771
54. Wei Y, Yu F, Song M, Han Z (2019) Joint optimization of caching, computing, and radio resources for fog-enabled IoT using natural actor-critic deep reinforcement learning. IEEE Internet Things J 6(2):2061–2073
55. Wang C, Liang C, Yu FR, Chen Q, Tang L (2017) Computation offloading and resource allocation in wireless cellular networks with mobile edge computing. IEEE Trans Wirel Commun 16(8):4924–4938
56. Mao Y, Zhang J, Song SH, Letaief KB (2016) Power-delay tradeoff in multi-user mobile-edge computing systems. In: IEEE GLOBECOM'16, Washington, DC
57. Zhang X, Yang F (2017) Joint bandwidth and power allocation for energy efficiency optimization over heterogeneous LTE/WiFi multi-homing networks. In: IEEE WCNC'17, San Francisco, CA
58. Kwak J, Kim Y, Lee J, Chong S (2015) DREAM: dynamic resource and task allocation for energy minimization in mobile cloud systems. IEEE J Sel Areas Commun 33(12):2510–2523
59. Yang L, Cao J, Tang S, Li T, Chan A (2012) A framework for partitioning and execution of data stream applications in mobile cloud computing. In: IEEE International Conference on Cloud Computing, pp 794–802
60. Sheng M, Zhai D, Wang X, Li Y, Shi Y, Li J (2017) Intelligent energy and traffic coordination for green cellular networks with hybrid energy supply. IEEE Trans Veh Technol 66(2):1631–1646
61. Grant M, Boyd S, Ye Y (2009) CVX: Matlab software for disciplined convex programming. Available [Online]: http://cvxr.com/cvx/
62. IEEE P802.11 Wireless LANs (2004) TGn channel models. IEEE 802.11-03/940r4, Technical Report
63. Zhai D, Sheng M, Wang X, Li Y, Song J, Li J (2016) Rate and energy maximization in SCMA networks with wireless information and power transfer. IEEE Commun Lett 20(2):360–363

Chapter 6
Fog-Enabled Intelligent Transportation System

6.1 Introduction

Currently, transportation systems are an indispensable part of human activities. As people become more dependent on transportation systems, the transportation systems themselves are facing not only many opportunities, but also lots of challenges. First, traffic congestion has become a noteworthy worldwide issue as the number of vehicles has grown dramatically over the past decades. Congestion can further lead to an increase in fuel consumption, air pollution, and difficulties in implementing plans for public transportation [1]. Second, accident risks increase with the expansion of transportation systems. Every year, about 1.24 million people are killed in car accidents worldwide [2]. Undoubtedly, there is a need to reduce traffic accidents and to detect accidents once they have occurred to minimize their impact. In addition, how to effectively manage the transportation system, provide drivers with convenient information services, and support the ultimate autonomous driving, these are all issues that the intelligent transportation system (ITS) needs to be considered.

Foreseeing importance of intelligent transportation system, various countries have launched their research programs on smart transportation systems [3].

United States

In the United States, electronic route guidance system (EGRS) was the initial stage of ITS in 1970s. It is a destination-oriented system which can provide highway guiding for motorists at intersections by exchanging messages between vehicle and roadside [4]. In 1991, United States' congress enacted integrated surface transportation efficiency programs (ISTEA). Then, TEA-21 (Transportation Equity Act for the twenty-first Century) was formulated in 1998 as a successor of project ISTEA. During this bill's covering period, 1997–2003, a large amount Federal funding for highway was authorized, and the vehicle infrastructure integration (VII) was also proposed. The VII provides a communication link between a vehicle to

Y. Yang et al., *Fog-Enabled Intelligent IoT Systems*,
https://doi.org/10.1007/978-3-030-23185-9_6

vehicle, and also to the roadside infrastructure. The key technology is emphasized on the well-known dedicated short range communication (DSRC), which will be mentioned later again.

On December 8, 2009, United States Department of Transportation (USDOT) started its 5 year Strategic Plan, which defines the strategic direction for the USDOT's ITS research program for the next 5 years [5]. 2018–2022 is the third 5-year phase, reflecting the four strategic goals of: safety, infrastructure, innovation, accountability [6].

European Union

European countries began to seriously study ITS technology in 1980s, with representatives of UK, France, and Germany. PROMETHEUS (Programs for European Traffic with Highest Efficiency and Unprecedented Safety) is believed to be the first research program, and DRIVE (Dedicated Road Infrastructure for Vehicle Safety in Europe), the second phase of Europe's R&D part of the framework, in 1988.

In 1991, the EU further promoted the study of ITS and set up ERTICO, which is a non-profit cooperative organization to strengthen cooperation between government and private enterprises. CVIS is attributed to be a successful outcome. CVIS (cooperative vehicle-infrastructure systems) is supposed to make cars be able to communicate with each other and know the nearby roadside infrastructure [7].

Currently, cooperative-ITS (C-ITS) and its evolution to support full autonomous driving is the motivation of research. The project COOPERS (Cooperative Systems for Intelligent Road Safety) plans to connect vehicles on the motorway to the road infrastructure via continuous bidirectional wireless communication.

Japan

Japan has got impressive benefits from investment in ITS and applying novel technologies into the operational deployment. Japan began their research in 1996 and created the world's first vehicle information communications system (VICS) which has been available nationwide since 2003. VICS provides real-time traffic information to vehicles in three technical ways, radio beacon, infrared beacon, FM multi-frequency broadcast, covering the entire city road [8].

UTMS'21 (universal traffic management system) is Japan's intelligent traffic management system with continuous improvement. As one of the most advanced ITS in the world, it can fully manage the traffic flow, committing to achieve a safe, comfortable, and environmentally friendly transport society [9].

Others

As the largest developing country, China is facing serious traffic problems in its development. Within some cities, China started its practice of intelligent traffic management, intelligent public transportation system, and traffic information service system since 1990s[10, 11]. In the "Made in China 2025" plan released by the State Council in 2015, the overall planning and promotion of R&D of smart vehicles are mentioned. Next stage, the use of novel intelligent and information technology like big data analytics and cloud computing throughout the transport system will be focused.

In December 2000, South Korea ventilated a 20-year blueprint for ITS development called National ITS Master Plan for the twenty-first century which singles out seven potential application areas. The scope of the plan ranges from local to national level to mitigate both immediate traffic problems and projected future mobility threats from the long-term perspective [12]. The plan is divided into three phases, and currently it is in the third phase of 2011–2020. This phase is to ensure the system type connection, compatibility, and efficiency of operation and to plan more advanced technology systems for the advanced stage. For now, this plan has basically been completed.

Enabled by fog computing, ITS data and information can be processed not only in cloud or on-board but also at any place in the continuum from vehicles to cloud. It saves huge bandwidth as well as introduces much shorter time delay. This chapter introduces architecture and key technologies of fog-enabled ITS and lists a number of use cases.

6.2 Intelligent Transportation System

Generally, ITS is recognized as using information, communication, control, computer technology, and other current technologies to establish a real-time, accurate, and efficient transportation management system. Figure 6.1 from ETSI provides a classical overview of the system concept: connecting vehicles, roadside (traffic) infrastructure, and central infrastructure to improve safety and traffic efficiency on roads.

6.2.1 Architecture and Key Components

Intelligent transportation systems (ITSs) are emerging as an effective means of both reducing traffic congestion and enhancing transportation safety[13]. ITS is an integrated system of people, roads, and vehicles, utilizing a variety of advanced technologies in communication, automation, computing, etc. An ITS uses the data collected from various sources, such as cameras, sensors, global positioning system (GPS) receivers, and other vehicles, to optimize the system's performance in terms of traffic flow, safety, delay, and fuel consumption.

The core components of an ITS are ubiquitous road environmental sensing and a vehicular communication system [14]. The sensing information, such as road conditions, driving status, and traffic information, is processed and shared by vehicles and RSUs. The communications medium among vehicles and RSUs is based on radio frequency technologies specifically designed for vehicular communication.

In the vehicular communication system, vehicles and RSUs share transportation sensing information via messages, as shown in Fig. 6.2. The messages may be sent periodically or on-demand, transmitted in one or multiple hops. The vehicles

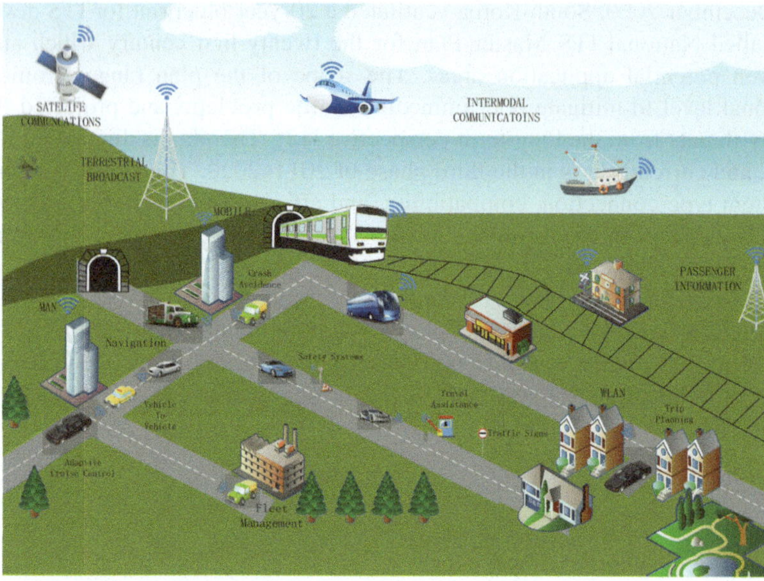

Fig. 6.1 ETSI's ITS scenario overview

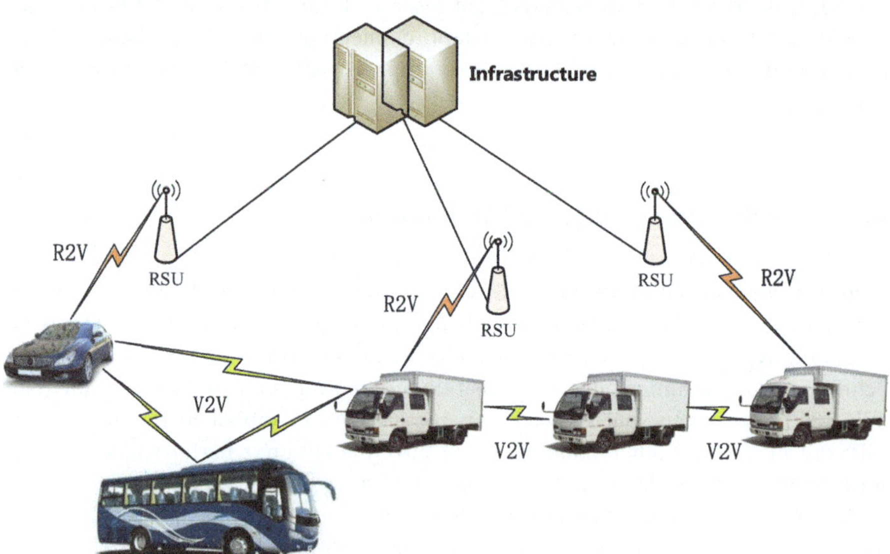

Fig. 6.2 Vehicular communication system

usually use messages to share information about themselves, such as location, speed, direction, and road condition events. RSUs send broadcast warning messages mainly about road condition and environmental hazards, and sometimes provide services to the vehicles. Such communication is commonly referred to as vehicle-to-vehicle (V2V) communication or RSU-to-vehicle (R2V) communication.

6.2.2 Vehicle Categories and Requirements

ITS is a broad ecosystem that includes personal vehicles and commercial vehicles that are on the roads as stand-alone vehicles or as vehicle fleets. There would be much benefit to define the norm of the ecosystem clearly before we taking deeper insight. Thus, several vehicles categories with different requirements are shown below.

- **Personal Vehicles**: Personal vehicles are pioneers in the evolution of ITS. Currently, some personal vehicles are already integrating some automatic driver assistant services (ADAS), and are actively evolving to more advanced automation levels.

  As vehicles become increasingly autonomous, they require

  - Additional sensing capabilities to increase safety;
  - Continuous awareness of vehicle and environmental status delivered to passengers;
  - Regional services provision, such as maps services;
  - Enhanced passenger comfort through contextual information, such as augmented reality (AR);
  - Assisted computing offered by the road infrastructure and other devices.

- **Commercial Vehicles**: Commercial vehicles include city buses, delivery vehicles, long-haul trucks, employee shuttles, tour company buses, and so on.

  The requirements of commercial vehicles include:

  - Continuous connectivity;
  - Coordinated driving for mobile platoon;
  - Multiple radio access technology connectivity to match different purposes ranging from supporting passenger productivity to enhancing tourist experience.

- **Truck Fleets**: Truck fleets can be seen as a special case of commercial vehicles. The fleets could be composed of trucks from different companies having different makes and models. The fleet operators want to closely monitor driving conditions, fuel efficiency, overall scheduling of goods to be transported, and route planning and optimization.

The requirements of truck fleets include:

- Road condition monitoring, such as receiving advance notification of road repair, congestion or accident;
- Detection of anomalous driving patterns, such as indication of stolen trucks;
- Truck condition monitoring and driver health monitoring;
- Maximizing fuel efficiently via convoying;
- Driving through regions without cellular connection.

6.3 Current Solutions and Technical Challenges

As described in Chap. 3, the challenges of ITS mainly focus on two aspects: computing and communication, while security and interoperability are the prerequisites of the system. Followings are some exiting solutions to these challenges.

6.3.1 On-Vehicle Computing

At present, there are a large number of computers or microprocessors in the automobile. Unlike that in consumer electronics devices, vehicular chips are working in aggressive environment with vibration, electromagnetic interference, and extreme temperatures. Related to the personal safety of passengers, it responds to higher reliability requirements.

Vehicular chips are designed for specific functions and reliability guarantees, such as error detection, authentication, etc. Qualcomm, NVIDIA, Intel, and so on are marking capable chips for coming smart driving. Due to the high requirements and complexity of automotive chips, in the face of rapidly growing intelligent automotive applications, the production speed of chip hardware is becoming a constraint. At the same time, the manufacturers are building their own camps and have formed a multi-standing situation, resulting in a variety of incompatible standards, which hindered the free development of smart vehicle applications.

In terms of in-vehicle networking, the bus network has been in the industry to maintain a solid position, over the past few decades of development. Multiple electronic control units (ECU) are connected by multiple in-vehicle LANs differing in transmission speed and communication protocol, according to the features and characteristics required for each application, exchanging information and coordinating control to allow more added value functions to be implemented [15]. Currently, common bus standards include but are not limited to:

- CAN (Controller Area Network): Developed by Bosch, the leading standard for over 20 years of automotive networks, it has a high level of safety and reliability for the control of key components such as ABS, auxiliary driving, engines, transmissions, and more.

- LIN (Local Interconnect Network): Master–slave structure, using polling methods, commonly used in windows, seats, lighting, and other less real-time demanding control.
- MOST (Media Oriented Systems Transport): The MOST bus networking is responsible for vehicle media system control data and the exchange of media data. It can be seen as a set of systems independent of driving control.
- FlexRay: FlexRay is a high-speed communication protocol that provides a high degree of flexibility and reliability. It is the basis for active technology development in Japan and worldwide. It uses time-triggered media access control and strict message delivery cycle, mainly used for wire brake, remote control throttle, and other emergency systems.

Figure 6.3 shows the architecture of the bus network in a present-day motor vehicle.

Towards the Internet of vehicles era, the above-mentioned vehicular network is still the basis of smart car or vehicle station. Facing with the need for intelligent control and the Internet of cars, engineers naturally think of using existing platforms to mount more hardware on the bus. For example, a transceiver can be used to connect to other vehicles or road side infrastructures, and a powerful computer is introduced to perform visual, location, and other data processing to realize assisted driving. But as a result, complexity and security are worrying. But in this way, reliability and security are fraught with complexity. Manufacturers are keen to

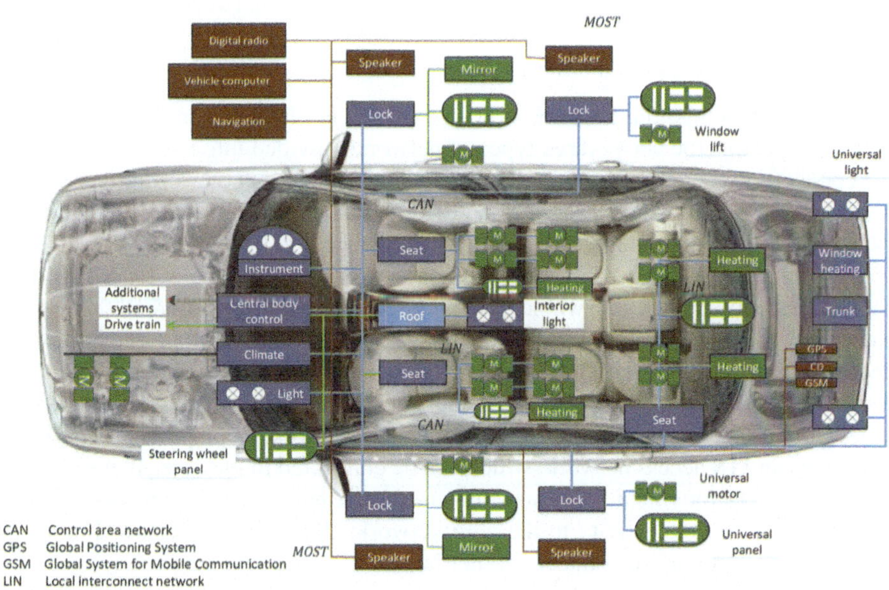

Fig. 6.3 Modern vehicle's network architecture

develop their own dedicated systems to provide vehicle connectivity control, but vehicle-mounted mobile communication data connections are considered to be the most secure and largest external interface. At Black Hat 2015, they demonstrated the remote control of a Jeep Cherokee car with a laptop via an "0day" vulnerability attacking onboard entertainment systems [16]. Some researchers also invaded a simulated vehicle system and gained fatal authority[17].

The vehicle nodes supported by the fog calculation solve these problems, and we will see below how the vehicle nodes can provide a unified and secure service using emerging technologies.

6.3.2 Communication Network

In ITS, besides the transmission of external information (such as Internet information), it is more common to employ a traffic-specific communication system for the communication of critical information to ensure its performance. Therefore, ITS's communication network architecture is a combination of external domain and ITS internal domain. In addition, ad hoc networking between vehicles is a promising form, which is also worth mentioning. So, in the form of ITS networking, there are three types defined as:

- ITS ad hoc network
- Access network (ITS access network, public access network, private access network)
- Core network (e.g., the Internet)

According to this, the network structure of ITS is illustrated in Fig. 6.4. In the figure, it contains the above three types of networks, divided into two domains, and the connectivity between the network is also indicated. It is noteworthy that in the traditional ITS architecture, services and resources are centrally deployed, which means that most of the vehicle applications need access to the core network.

Be sure, in most cases not all of these networks have to be accessed, and a part of the networks in the whole ITS architecture make up different scenarios, basically. According to ETSI's standard, they are grouped into four categories of deployment scenarios [18]. These scenarios are shown in Figs. 6.5, 6.6, 6.7, and 6.8, respectively: Deployment scenario A establishes a ITS ad hoc network, which can access to the core network by means of ITS access network; Deployment scenario B represents an ITS access network, which can be connected to the core network (e.g., the Internet). Deployment scenario C is based on a public access network, which can also provide connectivity to the core network. Deployment scenario D uses a private access network to connect to other networks or the core network.

Corresponding to the above architecture, the protocol stack of a single ITS station is shown in Fig. 6.9. This protocol stack consists of four horizontal layers: access technologies, networking and transport, facilities, and applications. In addition, it is flanked by a management layer and a security layer [19].

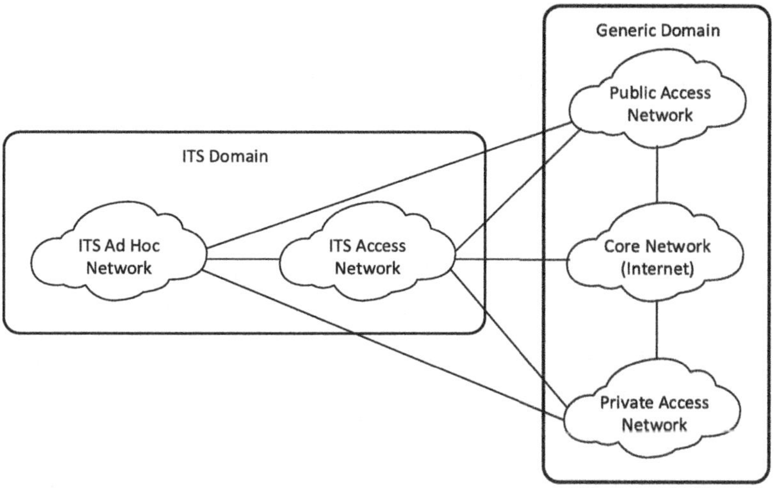

Fig. 6.4 Networks involved in the ITS architecture and their connectivity

Fig. 6.5 Scenario A: Ad hoc-centric

Fig. 6.6 Scenario B: ITS access network-centric

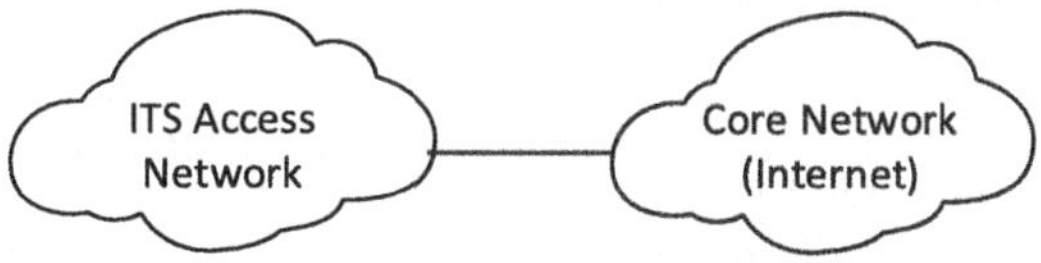

Fig. 6.7 Scenario C: Public access network-centric

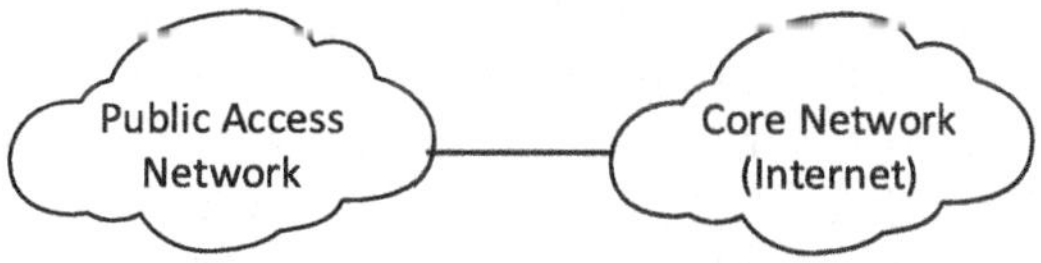

Fig. 6.8 Scenario D: Private access network-centric

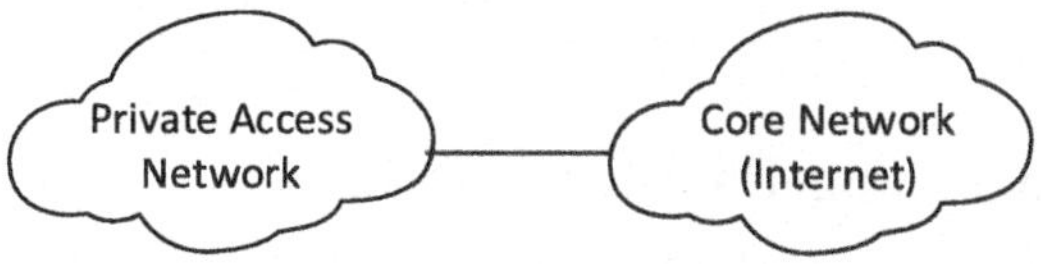

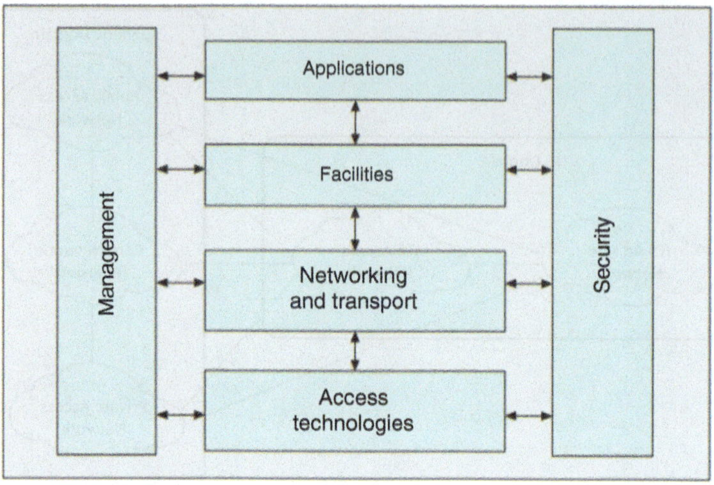

Fig. 6.9 Protocol stack of an ITS station

Till this moment, several organizations in different regions have defined the network standards for ITS. In October 1999, Federal Communications Commission (FCC) approved the frequency of 5.9 GHz (from 5850 to 5925 MHz) as the dedicated frequency for the intelligent transportation services based on DSRC technology, which is a milestone in the field of ITS [20, 21]. The standardization organizations including Institute of Electrical and Electronics Engineers (IEEE) and Society of Automotive Engineers (SAE) have jointly published a series of DSRC standards for V2V and R2V, such as IEEE 802.11p, IEEE 1609 series, SAE J2735 and J2945 standards, and the communication protocol of DSRC is shown in Fig. 6.10. In 2009, the European Commission authorized ETSI, European Committee for Standardization (CEN), and European Committee for Electro technical Standardization (CENELEC) to formulate a series of standards for intelligent transportation. In February 2014, ETSI and CEN jointly released the first version of the standard, which includes the ITS-G5 standard for safe driving and the emergency call (eCall) standard that provides emergency call services over cellular networks.

With the development of intelligent transportation technologies, the scope of ITS will be further expanded to the field of AD, smart navigation, automotive information and entertainment services, and other security services, such as collision warning and remote vehicle condition diagnosis.

6.4 Fog-Enabled Solution

Fog computing provides a critical architecture for today's connected world as it enables slow latency, high reliable, and high efficient operations, as well as provides strong support for mobile applications.

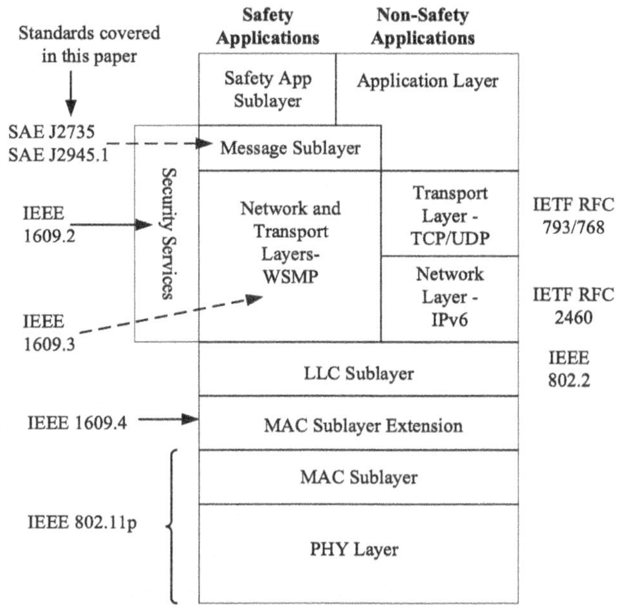

Fig. 6.10 Seven-layer DSRC communication protocol

Fog computing is an extension of the traditional cloud-based computing model. Fog computing exploits the computing, storage, communication, and control functions of the devices from cloud to network edge until terminal, which enables services to be distributed anywhere along the continuum between cloud and things. By flexibly allocating and managing the various resources in the cloud-to-thing continuum, fog computing can efficiently and intelligently provide services for multiple vertical industries and applications domains.

Based on the above features, fog computing enables the critical functions of ITS by collaborating, cooperating, and utilizing the resources of underlying infrastructures within roads, smart highways, and smart cities. Fog computing will address the technical challenges in ITS and will help scale the deployment environment for billions of personal and commercial smart vehicles.

6.4.1 Architecture and Key Technologies

In the hierarchical architecture as shown in Fig. 6.11, multiple FNs are distributed in vehicle, RSU, infrastructures, and the cloud, which provides a computing platform for ITS. Several different communication technologies, including wireless (e.g., 3G,

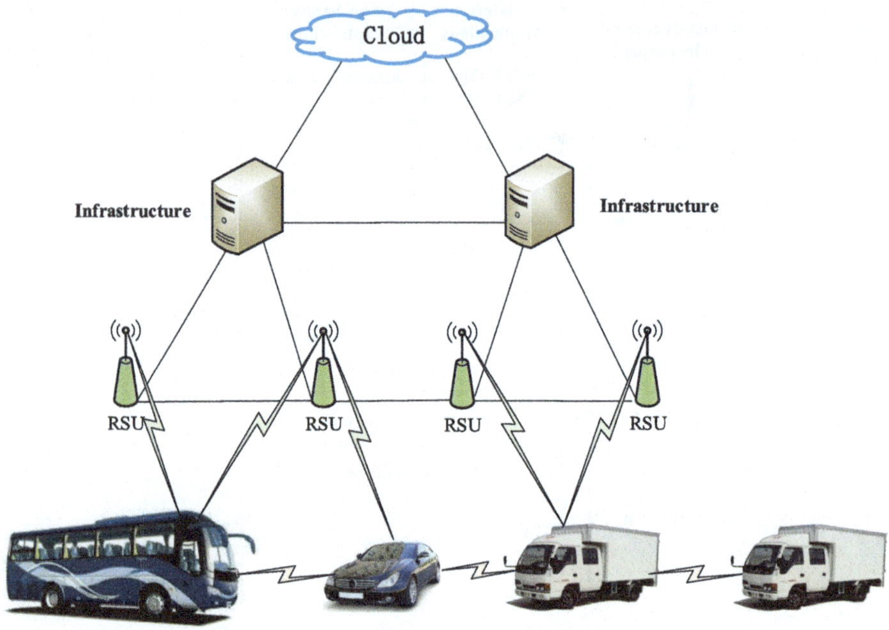

Fig. 6.11 Network infrastructure of connected vehicles

4G, 5G, DSRC, etc.) or wired (e.g., cable or optical fiber) technologies, can be used to provide secure inter-vehicle connectivity as well as the connectivity between vehicles, RSUs, infrastructures, and the cloud(s).

The features and functions of FNs in the hierarchical architecture are described as follows:

- **In-vehicle FN**
  In-vehicle FNs take over a series of sensors, e.g., radar, LIDAR, camera, GPS, gravity sensor, etc., which collect a massive amount of data for vehicle-to-everything (V2X) interactions. Their main role is to continuously sense the status of the vehicle and environment and to adapt the driving behavior by performing complex analytics on the available data in order to identify the best course of action. These analytics can be done by local computing processors, or with the assist of central cloud, infrastructures, or RSUs. In addition to internal sensors, the in-vehicle FN also extends its sensing capabilities through messages from other vehicles or from RSUs.
- **FN in RSU**
  FNs in RSUs with sensor capabilities can provide useful regional information to vehicles:
  - Serve as data aggregation points to analyze, simplify, and extract information in raw data from vehicles;

  - Detect metrics such as vehicle density and ramp length;
  - Connect to roadside sensors and cameras that monitor the road conditions. And then the local analytics in the RSUs could analyze this data to determine route and traffic guidance to communicate to nearby FNs and the cloud;
  - Act as a relay between vehicles for cooperative driving.

- **FN in infrastructure**
  FNs in infrastructures enable a variety of additional functionality, which includes:

  - Provide QoS through prioritization of variety of data being routed through it to the cloud;
  - Provide secure communication through authentication and encryption;
  - Provide location-based services;
  - Provide significant regional computational capabilities, which can be used for tasks such as coordinating regional traffic patterns, optimizing smart highway efficiency, and looking for safety or security problems in lower level FNs.

- **Cloud**
  Cloud platforms are typically hosted in data centers and usually provide long-term storage for traffic from vehicles, RSUs, and other fog devices. In addition, cloud will build and train machine learning models that could be used for detecting safety patterns and acting on them. Cloud platforms also provide centralized services to all vehicles and devices, such as application lifecycle management, configuration management, security, and software updates.

The fog architecture is based on eight pillars, including security, scalability, openness, autonomy, RAS, agility, hierarchy, and programmability [22]. These pillars can address the technical challenges and provide strong support for ITS as shown below:

- **Security**
  Fog requires every FN in the cloud-to-thing continuum to have high-assurance security mechanisms. It also specifies the process by which fog computing manages the physical interfaces, wireless protocols, and packets and data being transferred, which ensures that the right packet gets to the right location. Meanwhile, fog computing requires the FNs within the same service domain to execute a common set of information security and privacy policies, in order to form a distributed trusted computing platform. This distributed platform can provide on-demand security services to resource-constrained devices as well as offer trustworthy information processing, storage, and transport throughout the fog-enabled ITS.
- **Scalability**
  This breakdown of operations between in-vehicles, RSUs, infrastructures, and cloud eases scalability, management, and orchestration. The scalability of fog architecture includes the hardware or software adaptation of individual FNs, the addition of FNs in the fog network, and the storage, network, and other services

scaled with the fog infrastructure. Management and orchestration can ensure the maintenance of service level agreements (SLAs) for transactions. Transaction is treated as a unit for the purposes of satisfying a request between the consumer and the provider. The transaction is managed via an orchestrator which understands the soft-bounds of both consumer/provider and subcontractor service. SLAs direct the behavior of the subcontractors to achieve the consumer/provider contractual agreements within the predicted boundaries set in the SLA.

- **Openness**
  The fog-enabled ITS considers compute and storage partition from vehicle-to-cloud to address bandwidth and latency requirements. Interoperability and data sharing take place at different phases and layers. In-vehicle FNs are used for sensors fusion, image processing, and local analytics to trigger autonomous, time-sensitive, and immediate action for tasks like path planning and safe driving. In-vehicle storage is used to cache data from local sensors, and also to cache data regarding hazards or route conditions communicated V2V. RSUs may gather processed data from vehicles to be further processed to enable information to be communicated to the cloud. Cloud provides deeper analytics for bigger patterns, route planning, and guidance. These capabilities must be open. Vehicle manufacturers, fleet owners, insurance companies, and government regulators all participate to establish an open multi-party ecosystem to improve their cost, functionality, safety, and rate of innovation.
- **Autonomy**
  Each FN from vehicle to cloud has certain capabilities of processing, storage, and decision-making. Vehicles and RSUs are autonomous and can perform critical functions without the assist of cloud resources. To reduce latency and cost, it is benefit to make decision and store data in the low layer of the fog hierarchy.
- **RAS**
  Having each vehicle, RSU with compute and storage, and access network infrastructure with compute and storage as a FN can ensure RAS in the ITS. By decreasing the batch size of updates required either at the RSU or infrastructure level, the risk of dropped connection can be decreased during a handover, which is vital for both safety and navigation purposes. In addition, a group of vehicles can act as federated resources for compute and path planning, which ensures RAS even if there is no connectivity to the access network (e.g., in rural areas or in a tunnel).
- **Agility**
  Each FN from vehicle to cloud can agilely make immediate decision. Pushing the dynamic location update schemes from the cloud to the lower layer FN will greatly increase the agility of the system by enabling real-time reactions to location-based traffic patterns or services provided. Agility allows the fog network to adapt quickly to changing conditions or changing customer needs.
- **Hierarchy**
  Smart vehicles require connectivity along the fog hierarchy to distribute computing workload and data storage. This takes place through:

  - Thing-to-Fog: sensors inside the vehicle connect to the in-vehicle FN, and sensors on roads connect to FN in RSU;
  - Fog-to-Fog: in-vehicle FN connects to an in-vehicle FN, in-vehicle FNs connect to FN in RSU or FN in infrastructure;
  - Fog-to-Cloud: in-vehicle FN connects to the cloud, FN in RSU connects to the cloud, and FN in infrastructure connects to the cloud.

  This hierarchy will greatly improve the paging resolution from the cloud, as a vehicle may be handed over between RSUs at a rapid pace, and those RSUs will connect to a single infrastructure that maintains a connection to the cloud.
- **Programmability**
  Programmability is at the heart of a fog-enabled ITS and pervades all operations involving computing, storage, communications, and layering through a fog hierarchy. In-vehicle FNs are sufficient for target autonomous processing to be performed within the vehicle. When additional processing is needed, the partially processed information is conveyed to a more capable FN in nearby RSU and infrastructure or to a cloud. Federation of computing, storage, and communication enables dynamic adjustment of platform capabilities depending on needs. In contrast, virtualization techniques such as network slicing enable creation of individual logical resource spaces customized to the resource needs of the application. Different constituencies may generate code that runs in the fog-enabled ITS, which includes vehicle manufacturers, service providers, fleet owners, insurance companies, third party customizers, vehicle owners, and vehicle drivers.

6.4.2 Virtualized and Intensive Vehicle Stations

For vehicle nodes, the opportunity to step into a more advanced intelligent transportation system is to (1) introduce the onboard Ethernet technology; (2) adopt fog-enabled vehicle nodes. Using Ethernet technology to replace the current bus network framework will bring obvious advantages in application cost, transmission rate, and technical reserve. And the utilization of virtualization technology is the key to promote the development of the onboard node. By building the vehicle station into a fog computing node, forming a unified on-vehicle platform manageable and interoperable, diversified applications can be easily deployed.

Virtualization is the main method to provide isolated environments in fog computing and also the main factor of FN performance [23]. In response to an intelligent transportation system, vehicles are increasingly equipped with computing and storage capabilities. How to manage and schedule these resources while providing security guarantee has become a new challenge. Virtualization technology virtualizes physical devices into unified and resilient resources of storage, computing, and communication, forming a standard software platform that facilitates software development and deployment. Data between different services is isolated from each other, thereby increasing security. Virtualization technology also makes it possible for rapid migration, allowing a component to migrate to a more suitable physical location as needed during runtime. There are two main technologies for virtualization, conventional virtual machine, and container. All of them can realize resource virtualization on the physical machine. The difference lies in that the virtual machine simulates a completely identical environment with the physical bare metal. An operating system needs to be installed first and a running platform must be set up before the application can be deployed. Container technology such as Docker builds a lightweight virtual operating environment through software features, within the same operating system. Containers are especially well-suited for resource-constrained IoT hardware, taking the significant advantages of containerization in terms of flexibility and easy deployment [24, 25].

Furthermore, after the vehicle FN is popularized, vehicle fog computing (FVC) can be developed, and the concept of vehicle as a platform (VaaP) is proposed [26]. The proliferation of vehicle applications requires a greater demand for communications and computing infrastructure, without them, the Internet of vehicles and novel services cannot be put into practice, but only stay beyond the concept. The vehicle cloud can be set up to share the workload of the RSU by utilizing vehicles traveling on the road and parked in the parking lot as FNs, which provide with communications and computing resources for the whole. By discussing four types of scenarios of moving and parked vehicles as the communication and computational infrastructures, respectively, a quantitative analysis of the capacities of VFC is obtained that initially proved the feasibility of this idea.

6.4.3 Distributed Resources in Communication Networks

Fog computing aims to build a continuum between end users and the cloud to take full advantage of access network resources. From a perspective of communications network, we are going to discuss the uniqueness of fog computing in vehicular network architectures and the benefits of a fog-enabled network.

Vehicular ad hoc networks (VANETs) is a type of vehicle-specific ad hoc networking technologies in ITS that consists of interconnected vehicles. In recent years, VANET has attracted a lot of research, but because of limited connectivity and QoS guarantees, it is difficult to promote the deployment. The introduction of SDN technology and fog computing can enhance VANET performance, augmenting V2V, vehicle-to-infrastructure (V2I), vehicle-to-base station communications. As a result, a new SDN-based VANET architecture leveraging fog computing called FSDN VANET is proposed [27]. SDN separates the control plane from the data plane. The switch responsible for data forwarding does not need to be involved in complex rule decisions and therefore can be implemented with inexpensive, standardized equipment. OpenFlow is one example of such an SDN technology proposed by Stanford University. On the other hand, the fog computing feature moves resources down to the access network, providing low latency features, since

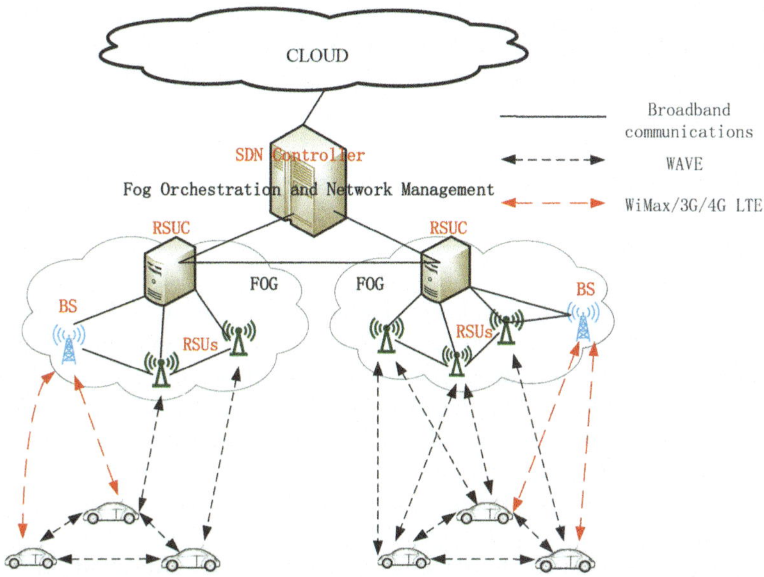

Fig. 6.12 SDN-based and fog-enabled VANET architecture

computing tasks can respond as close as possible to the data source. As shown in Fig. 6.12, the architecture of FSDN VANET consists of the following parts:

- **SDN Controller**: It has global intelligence and controls all the network behaviors of the entire SDN-based VANET system. It also plays as fog orchestration and resource management for the fog.
- **SDN Wireless Nodes**: The vehicles act as the end users as well as forwarding element, equipped with OBU and operating OpenFlow. They are data plane elements.
- **RSU**: RSU running OpenFlow and controlled by the SDN controller. It is a fog device.
- **Road-Side-Unit Controller (RSUC)**: A cluster of RSUs are connected to a RSUC through broadband connections before accessing to the SDN controller. RSUC is OpenFlow-enabled and controlled by SDN controller. Besides the responsibility of forwarding data, RSUCs also store local road system information and perform emergency services. RSUCs are fog devices, under the orchestration of SDN controller.
- **Cellular BS**: In this proposal, BS is not simply carrying voice calls and conveying data, it is more sophisticated. BS is under the control of SDN controller, running OpenFlow, capable of delivering fog services. Similar to RSUC, BS is also a fog device under the control of SDN controller with local intelligence.

In short, the fog-enabled network is a distributed, heterogeneous, and horizontal architecture. In the future, with the booming of smart car business and the surging of data transmission volume, it is foreseeable that as a viable solution, fog-similar network architectures will be the trend. How to manage and orchestrate resources, design an SDN controller suitable for telematics, as well as fall-back/backup mechanisms in the event of a failure remain to be the research challenges in the field of fog-enabled vehicle network.

6.4.4 Use Cases and Application Scenarios

Autonomous Driving

AD technologies are developing faster than anyone could possibly imagine. AD vehicles bring a whole new ecosystem with new requirements on the network architecture to support the huge amount of data processing workload and to satisfy the real-time services requirements. For AD, precise operations during every millisecond of driving time can have life-and-death consequence, which makes AD become a mission-critical application.

AD vehicle is dependent on both sensing and large amounts of software, such as situation awareness, route planning, vehicle control, etc. The sensing system and the software can cooperate with each other to realize a whole AD application.

In the fog-enabled ITS, the AD vehicle is a mobile FN that can communicate with the FNs in other vehicles, RSUs, or infrastructures. However, it must be capable of performing all required in-vehicle operations autonomously if it cannot connect to other FNs or the cloud.

The in-vehicle FN can directly communicate with the RSUs. The RSUs provide road condition and traffic data that the mobile FN uses to make routing and driving decisions, as well as deal with road conditions that it has not yet encountered from its localized sensors (e.g., water, snow, ice, lane closures, etc.). The in-vehicle FNs and FNs in RSUs allow multi-radio access to the vehicles to ensure continuous connectivity and to support services access. The FNs in infrastructures will provide road condition information to other traffic system FNs as well as to vehicles. And the in-vehicle FNs may also connect to other cloud systems, such as one or more service providers or government agency, in order to have the information available for the community of cooperating FNs.

On the other hand, as vehicles become increasingly autonomous, the data volume and computational complexity grow by an order of magnitude. In order to support the computational needs of the complex system, each vehicle needs distributed computing architecture with connectivity between multiple FNs supporting analytics, storage, and other resources. By distributing computation, the fog architecture can distill huge amounts of data generated by things closer to the data source. FNs can also combine multiple incoming streams of data through aggregation, compression, and others forms of processing and analyses, in order to assist the AD operations.

In summary, by applying fog computing into AD system, the following features are available:

- Extract information at a lower level in the fog hierarchy to reduce latency;
- Conserve limited and expensive resources, e.g., bandwidth, memory, and storage;

- Combine sources in different ways to meet the needs of different applications;
- Set and execute data transformation, data extraction, and data analytics closer to the data source to minimize the raw data transmitted to the cloud.

Cooperative Driving

Cooperative driving is important for fleets of vehicles that drive in a convoy and requires a convoy leader to constantly update the other vehicles with the direction, speed, and lane, road conditions, congestion, and other data. Cooperative driving is also useful for road alerts (e.g., hazards, traffic, or weather conditions), where the vehicles in the concerned area send alerts to other vehicles and also to RSUs for broader coverage.

In the fog-enabled ITS, the in-vehicle FNs in a convoy can establish a LAN to aggregate data traffic from multiple vehicles, and then transmit them over a cellular network, instead of each vehicle sending data to the fog or cloud via expensive cellular connection. This LAN is easy to set up as the vehicles in a fleet or convoy usually move at constant speed and constant distance between two vehicles.

Furthermore, a fleet of vehicles can act as federated resources for computing and path planning, which ensures reliability even if there is no connectivity to the access network. The in-vehicle FNs can collect a bunch of data from sensors in the convoy, and adapt the driving behavior based on variable road conditions and situations sensed locally.

On the other hand, when the convoy leader is unable to connect to the cloud, it can locate other FNs in the vicinity, which can provide continuous connectivity for fleet management. Since the computational workload is distributed across the fog infrastructure, including all the participating vehicles and nearby FNs, the convoy leader can make its own control decisions, even it is based on a more limited data set than the cloud might provide.

With fog computing, it enables a vehicle to leave the convoy and the convoy will self-heal. Conversely, a vehicle can also join the convey on the way. The convoy will be able to apply security mechanisms, such as discovery, authentication, and a reputation score to allow the new vehicle to join the convoy and place it in the correct position.

Shared Vehicles

Shared vehicle is a service that can complement the public transport. The users can easily access to shared vehicles in a broad range of areas, including apartment complexes, parking spaces, and large commercial districts without being limited to rental outlets. However, shared vehicles still have issues, such as how can users obtain the vehicle whenever and wherever they want.

The fog architecture for shared vehicles is shown in Fig. 6.13. The details are described as follows.

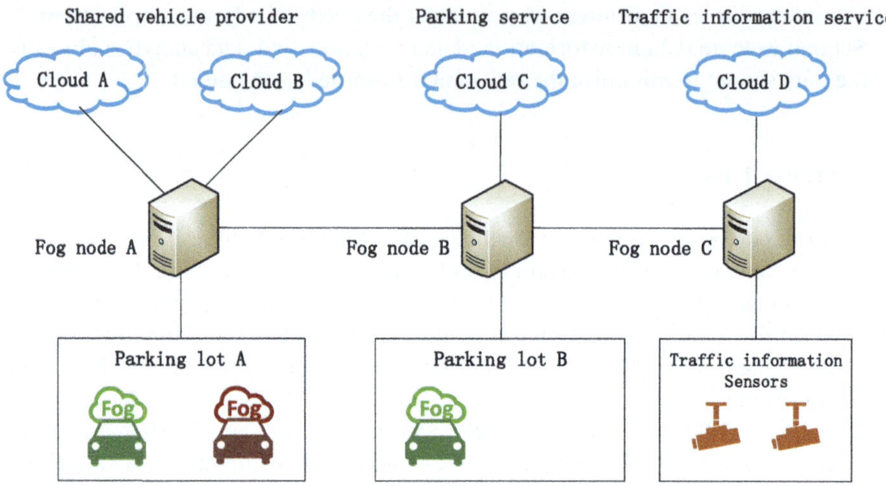

Fig. 6.13 Fog architecture for shared vehicles

When a user wants to rent a shared vehicle, he requests to the cloud A of company A with a specific request in parking lot location and time. The cloud A requests to FN A in parking lot A for finding vehicles matched to the request. If FN A does not find the matched vehicle, it sends the same request to other company's FNs. Cloud B of company B answers best matched vehicle. Then cloud A informs the user of the vehicle owned by company B. Cloud A also requests to the parking lot B near the destination of the user for finding unoccupied space. If there is unoccupied space, the user is scheduled to return the vehicle to parking lot B.

As soon as the user starts to drive the shared vehicle, the in-vehicle FN starts operating. The recorder begins recording video in case of rule violations, accidents, etc. The in-vehicle FN gathers real-time information from local sensors as well as other FNs (including other vehicles, RSUs, or FNs for traffic information service). The in-vehicle FN should react immediately when something may cause troubles.

If a shared vehicle causes an accident, the in-vehicle FN sends the video record to nearby FN. The FN generates accident information from this record and sends it to cloud A. Cloud A analyzes the accident information and arranges a tow truck and another vehicle with an insurance agent. The tow truck then takes the accident vehicle to a repair shop.

The benefits that fog computing provides to shared vehicles include:

- By data sharing among different companies, it is possible to immediately search for a vehicle that matches the user's requirements and a parking lot with unoccupied space near the destination;
- By constantly monitoring the vehicle condition and the real-time traffic information, the vehicles can react immediately to the emergency, which improves the user experience;

- When a traffic accident occurs, connections between FNs allow accident reports to be sent to the required points instantly and securely, which speeds up the accident handling.

6.5 Conclusion

Intelligent transportation system helps to improve traffic efficiency and ensure traffic safety. The core of this system is the collection and analysis of sensor data and vehicle communication technologies. Existing network architecture and communication technology still cannot meet the demand for advanced intelligent driving support and rapid development of intelligent transportation. As an emerging concept, fog computing is proposed for various IoT scenarios and can address the challenges in ITS. In this chapter, we first introduced the definition and development of ITS, describing the ecosystem composition and their respective requirements. Then, we explained the challenges and a stage-of-the-art of ITS, mainly focusing on vehicle station and communication network. To present fog computing, the architecture of fog-enabled ITS was provided. And we also discussed how fog computing can address the technical challenges and provide strong support for ITS. Finally, several use cases in fog-enabled ITS, including autonomous driving, cooperative driving, and shared vehicles, are shown in this chapter, which further verifies the benefits that fog computing can bring to ITS.

References

1. Shawe-Taylor J, De Bie T, Cristianini N (2006) Data mining, data fusion and information management. IEE Proc Intell Transp Syst 153(3):221–229
2. Al-Dweik AJ, Mayhew M, Muresan R, Ali SM, Shami A (2017) Using technology to make roads safer: adaptive speed limits for an intelligent transportation system. IEEE Veh Technol Mag 12(1):39–47
3. An S, Lee B, Shin D (2011) A survey of intelligent transportation systems. In: Third international conference on computational intelligence, communication systems and networks, Bali, pp. 332–337
4. Rosen DA, Mammano FJ, Favout R (1970) An electronic route-guidance system for highway vehicles. IEEE Trans Veh Technol 19(1):143–152
5. U.S. department of transportation. DOT strategic plan [eb/ol]. https://www.transportation.gov/dot-strategic-plan
6. U.S. department of transportation. DOT strategic plan for fy2018-2022 [eb/ol]. https://www.transportation.gov/sites/dot.gov/files/docs/mission/administrations/office-policy/304866/dot-strategic-plan-fy2018-2022508.pdf.
7. Verweij F (2005) Cooperative vehicle-infrastructure systems
8. Vehicle information and communication system (2013) How VICS works [eb/ol]. http://www.vics.or.jp/english/vics/index.html
9. UTMS society of Japan (2013) About UTMS: universal traffic management systems [eb/ol]. http://www.utms.or.jp/english/system/index.html

10. Lu HP, Li RM (2014) Developing trend of its and strategy suggestions. J Eng Stud 3(1):6–19
11. Ge Y, Liu X, Tang L, West DM (2017) Smart transportation in China and the United States. Center for Technology Innovation, Brookings Institution, Washington
12. Shah AA, Mahalik NP, Namkoong J, Lee JD (2006) Intelligent transportation-deployment and development process in Korea. WIT Trans Built Environ 22:89
13. Zhang J, Wang FY, Wang K, Lin WH, Xu X, Chen C (2011) Data-driven intelligent transportation systems: a survey. IEEE Trans Intell Transp Syst 12(4):1624–1639
14. Zhao M, Walker J, Wang CC (2013) Challenges and opportunities for securing intelligent transportation system. IEEE J Emerging Sel Top Circuits Syst 3(1):96–105
15. Renesas electronics corporation. In-vehicle networking solutions [eb/ol]. https://www.renesas.com/en-us/solutions/automotive/technology/networking.html
16. Miller C, Valasek C (2015) Remote exploitation of an unaltered passenger vehicle. Black Hat USA 2015
17. He Y, Gui He Q, Ming Hui S, Xin Y, Xuan Zhe W (2016) Cyber security and anomaly detection method for in-vehicle can. J Jilin Univ 46(4):1246–1253
18. ETSI. Intelligent transport systems (ITS) vehicular communications; GeoNetworking; part 3: Network architecture [eb/ol]
19. Kosch T, Kulp I, Bechler M, Strassberger M, Weyl B, Lasowski R (2009) Communication architecture for cooperative systems in Europe. Commun. Mag. IEEE 47(5):116–125
20. Wu X, Subramanian S, Guha R, White RG, Li J, Lu KW, Bucceri A, Zhang T (2013) Vehicular communications using DSRC: challenges, enhancements, and evolution. IEEE J Sel Areas Commun 31(9):399–408
21. Xu Q, Mak TK, Ko J, Sengupta R (2004) Vehicle-to-vehicle safety messaging in DSRC. In: International workshop on vehicular ad hoc networks, 2004, Philadelphia, PA, October, pp 19–28
22. OpenFog reference architecture for fog computing (2017) OpenFog Consortium
23. Dastjerdi AV, Gupta H, Calheiros RN, Ghosh SK, Buyya R (2016) Fog computing: Principles, architectures, and applications
24. Bellavista P, Zanni A (2017) Feasibility of fog computing deployment based on Docker containerization over raspberry pi. In: Proceedings of the 18th international conference on distributed computing and networking, ICDCN '17
25. Kaur K, Dhand T, Kumar N, Zeadally S (2017) Container-as-a-service at the edge: trade-off between energy efficiency and service availability at fog nano data centers. IEEE Wirel Commun 24(3):48–56
26. Hou X, Li Y, Chen M, Wu D, Jin D, Chen S (2016) Vehicular fog computing: a viewpoint of vehicles as the infrastructures. IEEE Trans Veh Technol 65(6):3860–3873
27. Truong NB, Lee GM, Ghamri-Doudane Y (2015) Software defined networking-based vehicular ad hoc network with fog computing. In: IFIP/IEEE international symposium on integrated network management, pp 1202–1207

Chapter 7
Fog-Enabled Smart Home and User Behavior Recognition

7.1 Introduction

One typical fog-enabled intelligent IoT system is the smart home, where each smart appliance/device is able to connect to the Internet and carry out some computing tasks. Each appliance/device can be viewed as an IoT node. These IoT nodes form a local network. In this scenario, to enable the home to better understand the humans and subsequently respond correctly, an efficient and secure human machine interact technology is necessary. Conventional remote controls are extremely inconvenient due to the larger number of appliances and the dependence on the hardware. A more efficient solution is to let the local network itself recognize the user behavior directly.

Human behavior recognition schemes can be classified into three categories, i.e., video-based, sensor-based, and radio-based [1]. Video-based behavior recognition uses cameras to capture the images or videos of humans and then extract the behavior information from those images or videos [2, 3]. These approaches can achieve high recognition accuracy. However, requirements for direct lines of sight and good lighting conditions have restricted the application of the method. Further, there are concerns about the privacy protection when cameras are utilized at home.

Sensor-based behavior recognition exploits the data obtained from the various ambient sensors, such as the gyroscope, the accelerometer, etc., to recognize behavior [4–6]. These approaches usually require users to wear the equipment and are commonly used to recognize large-scale activities, e.g., running and eating.

Radio-based behavior recognition uses the received radio signals reflected by humans to recognize different behaviors. The basic idea is that humans will scatter the radio wave and the movements of humans will make differences in the received signals. By analyzing the received radio signals, we can infer the information about the movements. These approaches allow the users to get rid of the constraints of hardware and will be able to protect privacy. Thus, radio-based behavior recognition has advantages in smart home scenarios where comforts and privacy protection are

Y. Yang et al., *Fog-Enabled Intelligent IoT Systems*,
https://doi.org/10.1007/978-3-030-23185-9_7

of our major concern. Meanwhile, numerous wireless communications between the IoT nodes in the smart home also facilitate the implementation of these approaches. In this chapter, we will mainly focus on this type of behavior recognition.

There have been many works studying user behavior recognition with radios. Considering the fact that the commercial Wi-Fi devices are ubiquitous, numerous researchers endeavor to make full use of the current Wi-Fi devices to recognize user behaviors. They extract the acquired CSI from the Wi-Fi chips with little changes to the Wi-Fi device drivers. Based on the extracted CSI, various behaviors can be recognized. A brief summary of existing researches in this area is listed as follows.

- **Activity Recognition**
  The concept of E-eyes [7] was proposed in 2014 which enabled recognizing both in-place activities and walking movements in home by examining Wi-Fi CSI features and matching the signal profiles. A CSI based human activity recognition and monitoring system [8] was proposed later in 2015.
- **Sleep Monitoring**
  Methods of monitoring the position changes [9] and respiration [10] of a sleeping people were respectively proposed in 2014 and 2016. A way of tracking vital signs of breathing and heart rates during sleep in both one-person and two-person scenarios [11] was carried out in 2015, after which in 2017 a deep learning framework [12] was applied to fulfill the task.
- **Fall Detection**
  An interesting proposal of contactless fall detection through Wi-Fi devices [13] was brought forward in 2014 and improved [13, 14] then in 2017.
- **Smoking Detection**
  Rhythmic patterns that smoke left on Wi-Fi signals could be used to detect smoke itself [15, 16], which was proposed in 2016.
- **Finger Movement Detection**
  The concept of WiGest [17] that changes in Wi-Fi signals strength could be leveraged to recognize hand gestures was proposed in 2015. Similar work named WiFinger [18] was carried out in 2017. The authors in [19] even used Wi-Fi devices to recognize keystrokes.
- **Motion Direction Inferring**
  WiDance [20] was proposed in 2017 which used Wi-Fi devices to infer motion direction and to realize a Wi-Fi based user interface for a dance game.
- **Tracking**
  A Wi-Fi based decimeter-level passive tracking and velocity monitoring system [21], namely Widar, was put forward in 2017. Its improved version, Widar 2.0 [22], was then proposed in 2018.

In addition to the off-the-shelf Wi-Fi, some researchers even developed new radio chips to recognize user behavior. One recent example is Google's Soli project [23, 24], which integrated a millimeter wave radar into a chip and designed the corresponding behavior recognition framework. Some researchers also tried to track the moving patterns to obtain quantitative input information [25, 26]. Note that, as the range resolution of the radar heavily depends on the wavelength, short wavelengths are preferable to achieve higher accuracy.

Besides, since the sound is also a kind of wave, some researchers took advantage of the acoustical signals to track the moving objects. The frequency of sound is much smaller than that of radio while they have the same wavelengths. The requirements on the hardware to process the sound wave is much less stringent. The speakers and microphones in cell phones can be employed to transmit or receive the sound signals [26, 27].

This chapter summarizes the up-to-date researches on user behavior recognition with radio waves, which will help gain insights into future smart homes when combining fog computing. The organization of this remainder is as follows. Section 7.2 provides a brief formulation of the problem. Section 7.3 summarizes the passive user behavior recognition. Section 7.4 introduces the active user behavior recognition and tracking. Section 7.5 discusses future directions.

7.2 User Behavior Recognition for Smart Home

In user behavior recognition, we usually predefine a set of behaviors, i.e., $\mathbb{A} = \{A_1, A_2, \ldots, A_P\}$, where P is the total number of predefined behaviors. As the user performs a specific behavior A_i, the corresponding radio signals picked up at the receiver can be written as

$$\boldsymbol{Y} = \{\boldsymbol{s}_1, \boldsymbol{s}_2, \cdots \boldsymbol{s}_M\}, \tag{7.1}$$

where $\boldsymbol{s}_i$ represents one raw data stream and M is the total number of data streams. For example, the receiver may receive M waveforms if it is equipped with M antennas. The raw data may be the channel state information or the received waveform in time domain. Our goal is to build a model G which can identify the performed behavior given the input data $\boldsymbol{y}$. In particular, the estimated behavior $\hat{A}$ is given by

$$\hat{A} = G(\boldsymbol{Y}). \tag{7.2}$$

In most cases, the raw data cannot be utilized directly. We need to extract useful features from the raw data. This procedure can be model as

$$\boldsymbol{F} = F(\boldsymbol{Y}). \tag{7.3}$$

Then the original behavior recognition problem can be modeled as

$$\hat{A} = G(F(\boldsymbol{Y})). \tag{7.4}$$

In general, the function F characterizes the aggregation of various signal processing tasks and the function G represents a certain learning algorithm.

A typical user behavior recognition framework includes the following four steps, i.e., base signal selection, preprocessing, feature extraction, and classification:

- Base signal selection:
  The raw data are acquired at the receiver. For Wi-Fi devices, CSIs are the raw data in most cases. For some specific hardware, such as the frequency-modulated continuous-wave (FMCW) radar or the Soli chip, the received baseband signals are the raw data. The useful information may lie in the amplitudes, phases, phase differences, or the other transformed signals.
- Preprocessing:
  The raw data need to be preprocessed before feature extraction. For example, the Hampel filter can be used to remove the outliers [9]. The finite impulse response (FIR) filters are utilized to remove the irrelevant information in the raw data. Further, the principal component analysis (PCA) is frequently applied to remove the redundancies in the raw data.
- Feature extraction:
  Appropriate transformations are generally necessary as features in multiple domains will be needed. One basic transformation is the Fast Fourier transform (FFT) which transforms the data from time domain to frequency domain. However, FFT is primarily applied to stationary signals and not capable of capturing the instantaneous frequency components caused by human dynamics. Therefore, short-time Fourier transform (STFT) is often exploited [20]. STFT is a window based Fourier and can capture instantaneous frequency components. Moreover, discrete Wavelet transform (DWT) can be used to increase the frequency resolution in some applications [17, 18]. The advantages of DWT over STFT are that: (1) DWT has nice tradeoff between time and frequency resolutions; (2) DWT groups frequencies that differ by several orders of magnitude into a few levels so that both high speed and low speed movements can be captured; (3) DWT can also reduce the data size to alleviate the compunction load. Various features need to be extracted in multiple domains. Common time domain features include standard deviation, average, correlation, range, and so on. Frequency domain features include main-power band, profile feature, and so on. A typical feature extraction example can be found in [23].
- Classification:
  Once the features are gathered, we can recognize the corresponding behavior based on the collected features. In general, there are two kinds of classifications, one is the rule based classification and the other is the machine learning classification. In some case, the feature differences between different behaviors are apparent and can be easily explained. Accordingly, a simple rule based classification works well [20]. Some works simply classify the data by comparing the data with a series of templates and choosing the best matched one. However, in most cases, the feature differences are not evident and the machine learning based classification is utilized to find the inner differences. Various machine

learning algorithms such as the support vector machine (SVM), the k-nearest neighbor algorithm (KNN), the hidden Markov model (HMM), the convolutional neural networks (CNN), the dynamic time warping (DTW), the Random Forest, etc. can be applied to classify the data. Some algorithms can achieve high recognition accuracy with high computation loads while the others are on the opposite. The choice of one classification algorithm is thus a tradeoff between the computing power and the desired recognition accuracy.

Typical radio-based behavior recognition can be classified into two categories according to the utilized raw data. One is the "passive approach" which takes the CSIs as the raw data. Many researchers adopt this approach in Wi-Fi framework because of the ubiquitous Wi-Fi devices and the convenience in obtaining the CSIs. The other is the "active approach." In this approach, we can get out of the constraints of the off-the-shelf commercial communication devices and transmit the signals specially designed for the particular recognition tasks.

7.3 Passive Approach

The passive approach is generally based on Wi-Fi devices. When Wi-Fi devices communicate with each other, CSIs are necessary to decode the received waveforms correctly. Since the CSI measures the effects of the surrounding scatterers on the radio waves, the dynamics in the CSI reflect the changes in the scatters. Thus we can use the CSI as the raw data to carry out user behavior recognition. The CSI information exists in commercial Wi-Fi chips and can be easily extracted out with little changes to the hardware drivers [28]. Thus a majority of user behavior recognition works are exploiting the Wi-Fi CSI framework. These approaches are passive as they reuse the raw data designed for other purposes instead of utilizing new data that are specially transmitted/received for the particular recognition tasks.

7.3.1 Signal Detection

Doppler Effect

The impacts of human behavior on the CSI are illustrated in Fig. 7.1. In the figure, multiple paths are received at the receiver. The human body acts as a set of scatterers. The human movements can be modeled as changes in the locations of those scatterers, which lead to changes in CSI at the receiver. The channel frequency response (CFR) at frequency f and time t results from the superimposition of the responses from all individual paths [8, 20], i.e.,

$$H(f,t) = \sum_k \alpha_k(t) e^{-j2\pi f \tau_k(t)}, \tag{7.5}$$

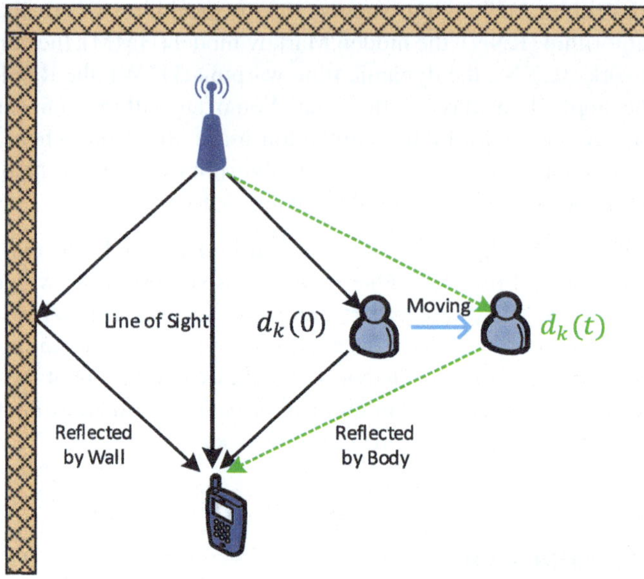

Fig. 7.1 An illustration of impacts of human moving on the CSI

where $\tau_k(t)$ is the delay of the k-th path and is determined by $\tau_k = d_k(t)/c$, $d_k(t)$ is the path length, and c represents the speed of light. The Doppler shift of the k-th path is

$$f_{D,k} = -f\frac{d\tau_k(t)}{dt} = -\frac{f}{c}\frac{dd_k(t)}{dt}. \tag{7.6}$$

Notice that some paths are not affected by human movements. Thus the channel frequency response can be divided into two parts, i.e., the static component and the dynamic component. In particular, we have

$$H(f,t) = H_s(f) + \sum_{k=1}^{P_d} \alpha_k(t)e^{-j2\pi f\tau_k(t)}, \tag{7.7}$$

where $H_s(f)$ is the static component which is the summation of those paths whose lengths are fixed during the analysis. Those paths include the light-of-sight (LOS) path and other paths reflected by the background static scatterers. The parameter P_d denotes the total number of dynamic paths whose lengths will change over time. Those paths are reflected by the moving scatterers. Different behaviors will lead to different changing patterns in the path lengths.

In practice, the estimated CSI at the receiver will suffer from phase shifts due to the carrier frequency offset (CFO) caused by hardware imperfection. The obtained CSI in the presence of CFO becomes

$$\bar{H}(f,t) = e^{-j2\pi \Delta f t} H(f,t) = e^{-j2\pi \Delta f t} \left(H_s(f) + \sum_{k=1}^{P_d} \alpha_k(t) e^{-j2\pi f \tau_k(t)} \right), \tag{7.8}$$

where Δf stands for the CFO.

Preprocessing

There are two steps in the preprocessing of the signal, including removing unknown phase shifts and useless signals. The two steps are explained respectively in this section. *Remove Unknown Phase Shifts:* Since the movement information is contained in $\tau_k(t)$ which is the phase of each path, we need to remove the unknown phase shifts due to the CFO. The following methods can be used:

- Signal Antenna: We can exploit the CSI power [8] as follows:

$$|\bar{H}(f,t)|^2 = |H_s(f)|^2 + \sum_{k=1}^{P_d} |\alpha_k(t)|^2 + 2\sum_{k=1}^{P_d} |H_s(f)\alpha_k(t)| \cos(2\pi f \tau_k(t))$$
$$+ \sum_{k=1}^{P_d} \sum_{i=1, i\neq k}^{P_d} 2|\alpha_k(t)\alpha_i(t)| \cos(2\pi f(\tau_k(t) - \tau_i(t))). \tag{7.9}$$

  Clearly the total CSI power is the sum of a constant and a set of sinusoids whose frequencies are functions of the path length changes or the user moving speeds. However, this method loses the sign information in the Doppler shifts which indicates the moving directions.
- Multiple Antennas: The unknown phase shift can be removed by multiplying the CSIs from two antennas [20]. In the case of multiple antennas, the CSIs of those antennas will have the same unknown phase shift. Thus multiplying the CSIs of two antennas can remove the unknown phase shift. Assuming $\bar{H}_1(f,t)$ and $\bar{H}_2(f,t)$ are the CSIs for two antennas, then we can have

$$\begin{aligned}\bar{H}_1(f,t)\bar{H}_2(f,t)^* - \; & H_{s,1}(f)H_{s,2}(f)^* \\ & + \sum_{k=1}^{P_d} \alpha_{k,1}(t) H_{s,2}(f)^* e^{-j2\pi f \tau_{k,1}(t)} \\ & + \sum_{k=1}^{P_d} H_{s,1}(f)\alpha_{k,2}(t)^* e^{j2\pi f \tau_{k,2}(t)} \\ & + \sum_{k,i} \alpha_{k,1}(t)\alpha_{i,2}(t)^* e^{j2\pi f \tau_{i,2}(t) - j2\pi f \tau_{k,1}(t)}. \end{aligned} \tag{7.10}$$

  The first term is a static component which can be removed with a high-pass filter. The last term is a cross term. The two terms in the middle are the target ones containing useful information. Note that the static component can dominate the dynamic components due to a strong LOS path or large static reflectors like the floor. Furthermore, the target terms are much stronger than the cross term and they include the Doppler shifts of interest. As the static component is almost constant and the dynamic component changes fast within a short analysis time window, we can roughly approximate the static component by the averaged CSI and approximate the relative strength of the dynamic components with the variance of the CSI.

Remove Useless Signals The received CSI contains lots of irrelevant signals. As a result, preprocessing is needed to obtain cleaner CSI. Since the Doppler shifts caused by human behavior lie within a limited bandwidth, FIR can be exploited to remove the useless signals. Further, note that Wi-Fi devices use orthogonal frequency-division multiplexing (OFDM) modulation and each subcarrier reports individual CSI. Thus the changes introduced in all the CSI streams by human behavior are strongly correlated. To make full use of the correlation while compressing the data, PCA is widely used to process the CSI stream. PCA of the CSI streams results in a series of principal components. A few principal components are retained and then passed to the following processing units. For example, the first principal component is retained in [20] while the first principal component is discarded and the next five principal components are retained in [8].

Feature Extraction and Recognition
After preprocessing the data, various features can be extracted. The features can be pruned according to some predefined rules or selected by some machine learning algorithms. Once we collect the features, the problem becomes a typical classification problem. Typical machine learning algorithm can be used in this phase. For example, SVM was used in [13], KNN was adopted in [19], and HMM was exploited in [8].

In summary, a passive approach typically makes use of commercially available Wi-Fi chips. The advantage lies in the low hardware cost. However, due to the centimeter-level wavelength of the transmitted radio signal, the passive approach can only recognize actions with relatively large moving ranges.

7.3.2 Wi-Fi Based Solution for Behavior Recognition

In 2017, Liu et al. from Tsinghua University reported their work on WiDance [20]. In particular, commercial Wi-Fi devices were utilized to implement a human-commuter interaction interface for a dancing game. The system can recognize eight moving directions of the player. See also Fig. 7.2.

The WiDance system includes one transmitter with a single antenna and two receivers with three antennas as illustrated in Fig. 7.3. First, during the preprocessing step, each receiver obtained the raw CSI from each receive antenna and the effect of CFO was eliminated by multiplying two carefully selected CSI streams from

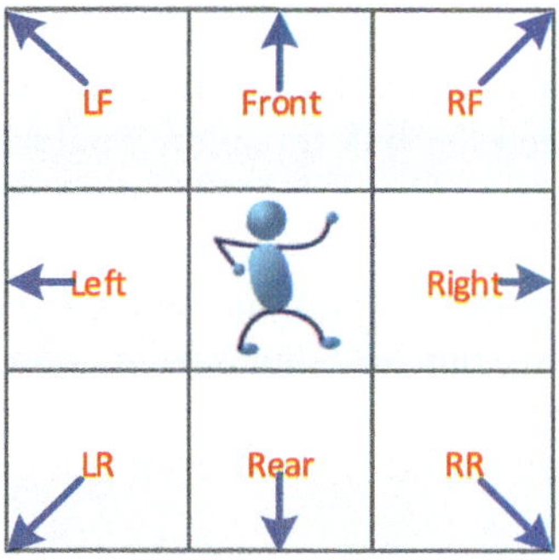

Fig. 7.2 Eight different moving patterns detected by WiDance

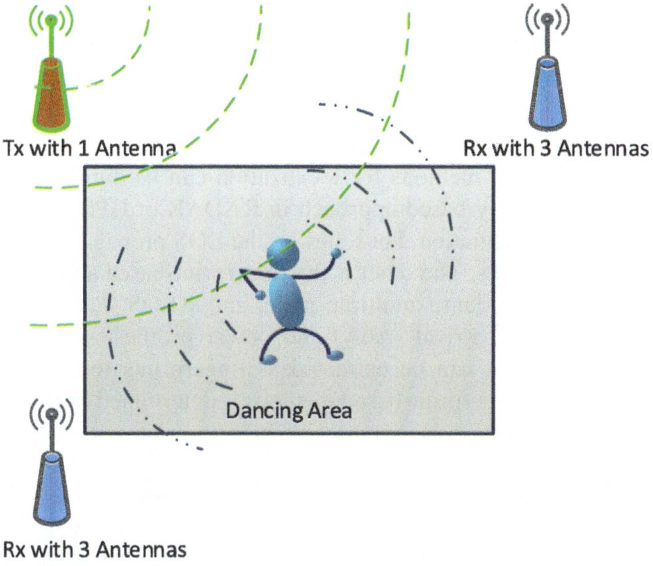

Fig. 7.3 The human–computer interface of WiDance

the three streams. A Butterworth passband filter was used for data sanitation. After that, PCA was applied to all CSI subcarriers and only the first principal component was preserved. Then an STFT with a Gaussian window was applied to get the spectrogram. Second, movement detection and trace segmentation were performed. The WiDance system calculated and smoothed the variations in the power distribution of the spectrogram and claimed there existed movements if the variance exceeded a predefined threshold. Since each behavior began with moving in one direction and then retracted back, each behavior resulted in a pair of peaks or valleys in the Doppler frequency shifts. Once a movement is detected, the system averages the absolute values of Doppler frequency shifts of the two links and takes the segment between two adjacent peaks as the spectrum of this action. Classification is carried out next. Unlike the widely used machine learning based classification, the WiDance system classified the behaviors according to the observed difference in the Doppler frequency shifts generated by difference behaviors. Specifically, the differences in the angles between the moving direction and the link direction would lead to differences in the amplitudes of the Doppler frequency shifts. Thus the behaviors could be classified into four groups according to the ratio between the accumulative absolute values of Doppler frequency shifts associated with the two links. Each group included two behaviors corresponding to two opposite moving directions. The two behaviors could be further identified based on the order of appearance of the negative and positive Doppler frequency shifts in the segment. An overall recognition accuracy of 92% was reported in [20].

7.3.3 Massive MIMO Based Behavior Detection and Ranging

With the growing demand of location-based services in wireless networks, e.g., indoor navigation, positioning has drawn increasing interest in both academia and industry. The fundamental methods for localization can be divided into two categories. One is the geometry-based approach in RADAR or GPS, which calculates locations via simple triangulation, but relies on the LOS propagation and the cooperation among multiple APs. The other is the fingerprint-based approach for indoor or urban conditions with dense multiple paths and NLOS. Particularly, channel features such as angle of arrival (AoA) and power profiles which characterizes the scattering environment can be extracted as fingerprints for localization. The positioning accuracy of fingerprint-based methods is determined by the performance of feature estimates. However, the complexity of appropriate extraction method is a severe burden [29].

Recently, massive multiple-input multiple-output (MIMO) as well as millimeter wave (mmWave) has been envisioned as the key technology in next generation wireless communications. By deploying a large antenna array at the AP and exploiting a wide bandwidth for transmission, the high resolution can be obtained in both angular and temporal domains, which provides feasibilities in improving the localization performance. In [30–33], the AoA, the received signal strength (RSS), or their combination was estimated from the large-scale mmWave channel to compute the location of the target user. The Cramer-Rao lower bound studied in these works proved that the positioning error could be reduced due to the channel sparsity in mmWave and the accurate angle estimation.

Above approaches mainly utilize the angle, delay, or signal strength knowledge at APs. However, additional information at the user side like the angle of departure (AoD) is rarely considered and can be regarded as another degree-of-freedom for localization. Meanwhile, in some applications such as indoor navigation, not only the user position but also its behavior, e.g., the moving velocity and direction, should be detected to predict and guide the user motion. Considering the benefits of massive MIMO and mmWave, we proposed a novel mechanism called *Massive MIMO based Detection and Ranging* (MIDAR) for joint localization and behavior

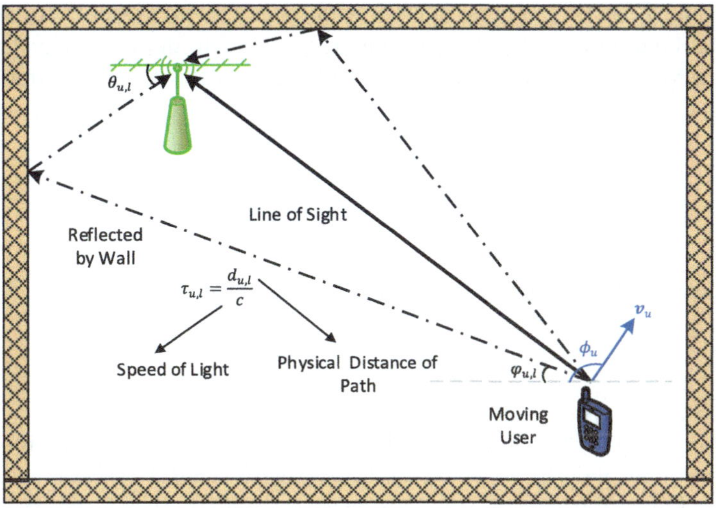

Fig. 7.4 Indoor channel model in MIDAR

recognition. Specifically, thanks to the superb angular and temporal resolution in mmWave massive MIMO, a direct but accurate enough approach for feature extraction was provided to obtain the multi-dimensional channel power profiles, i.e., the angle-delay-Doppler power spectrum (ADD-PS). Exploiting this ADD-PS as the 3-D fingerprint of a particular position with a certain behavior, MIDAR achieved joint localization and behavior recognition by fingerprint matching schemes or deep learning techniques. In the following parts, we give a brief overview on MIDAR.

Consider an indoor scenario where a mmWave AP located at a known position employs a large-scale antenna array of M antennas and serves several moving users with single-antenna. The user's location in the room is unknown and needs to be estimated. For indoor case, the uplink channel from user-u to the AP is the superposition of different propagation paths from L scatterers (e.g., surrounding walls, roof, or ground surfaces), known as multipath components. The l-th multipath component can be characterized by three position-related parameters: the delay $\tau_{u,l}$, the AoA $\theta_{u,l}$, and the AoD $\varphi_{u,l}$, as illustrated in Fig. 7.4.

Denote $\boldsymbol{h}_u[t, n] \in \mathbb{C}^{M\times 1}$ as the discrete-time UL channel vector of user-u at the n-th tap and in the t-th reference symbol (RS). Applying the wide-band OFDM system, it can be expressed as

$$\boldsymbol{h}_u[t, n] = \sum_{l=1}^{L} \alpha_{u,l} e^{j2\pi f_{u,l} t} \boldsymbol{a}(\theta_{u,l}) \delta \left(n - \frac{\tau_{u,l}}{T_s} \right), \tag{7.11}$$

where $\alpha_{u,l}$ represents the complex channel gain with zero mean and statistical power $\rho_{u,l}$, $f_{u,l}$ denotes the normalized Doppler frequency shift, $\boldsymbol{a}(\theta)$ stands for the array response, and T_s is the sampling interval within one OFDM symbol.[1] To avoid the inter-symbol interference, the delay should be less than the duration of cyclic prefix, i.e., $\tau_{u,l} < N_{\text{cp}} T_s$ where N_{cp} is the cyclic prefix length. Assuming a uniform linear array at the BS with antenna spacing d_c and carrier wavelength λ, $\boldsymbol{a}(\theta)$ has the following form when the AoA is θ:

$$\boldsymbol{a}(\theta) = \left[1, e^{-j2\pi \frac{d_c}{\lambda} \cos\theta}, \ldots, e^{-j2\pi \frac{d_c}{\lambda} \cos\theta (M-1)}\right]^T. \tag{7.12}$$

The normalized Doppler frequency shift $f_{u,l}$ is given by

$$f_{u,l} := \frac{\|\boldsymbol{v}_u\| \cos(\varphi_{u,l} - \phi_u)}{\lambda} \cdot \frac{1}{f_{\text{ch}}}, \tag{7.13}$$

where $\|\boldsymbol{v}_u\|$ denotes the magnitude value of velocity vector $\boldsymbol{v}_u$, ϕ_u represents the moving direction, and f_{ch} characterizes the density of channel observations in time.

By stacking $\boldsymbol{h}_u[t, n]$ of all taps into a matrix $\boldsymbol{H}_u[t]$ as $\boldsymbol{H}_u[t] := [\boldsymbol{h}_u[t, 0], \ldots, \boldsymbol{h}_u[t, N_{\text{cp}} - 1]]$, we can project the channel into the angle-delay domain as

$$\boldsymbol{G}_u[t] = \boldsymbol{F}_M^\dagger \boldsymbol{H}_u[t], \tag{7.14}$$

where $[\boldsymbol{F}_M]_{p,q} := e^{-j2\pi \frac{p(q-M/2)}{M}}/\sqrt{M}$, $p, q \in [0, M-1]$ is the M-dimensional phase-shifted DFT matrix. Then, collect $\boldsymbol{G}_u[t]$ during T successive RSs and define the temporal channel vector as $\boldsymbol{g}_{u,p,q} = [[\boldsymbol{G}_u[0]]_{p,q}, \ldots, [\boldsymbol{G}_u[T-1]]_{p,q}]^T$. The angle-delay-Doppler frequency domain channel can be further obtained as

$$[\mathscr{G}_u]_{p,q,s} = [\boldsymbol{F}_T \boldsymbol{g}_{u,p,q}]_s, \tag{7.15}$$

where $\mathscr{G}_u$ is the corresponding 3-D channel array, and $[\boldsymbol{F}_T]_{s,t} := e^{-j2\pi \frac{(s-T/2)t}{T}}/\sqrt{T}$, $s, t \in [0, T-1]$ stands for the T-dimensional phase-shifted DFT matrix. Since $\mathscr{G}_u \in \mathbb{C}^{M \times N_{\text{cp}} \times T}$ is actually a 3rd-order tensor comprised of channel gains in angle, delay, and Doppler frequency domains, algorithms regarding tensor algebra should be utilized for the feature extraction and further analysis.

Note that $\mathscr{G}_u$ still represents the instantaneous CSI. For fingerprint-based localization and behavior recognition, the wide-sense stationary characteristics determined by the stable scattering environment needs to be further extracted. Denote the ADD-PS $\mathscr{P}_u \in \mathbb{R}^{M \times N_{\text{cp}} \times T}$ as $[\mathscr{P}_u]_{p,q,s} = \frac{1}{MT} \mathrm{E}\left\{\left|[\mathscr{G}_u]_{p,q,s}\right|^2\right\}$. When M and T become large, it can be expressed as

---

[1]The sampled delay $\tau_{u,l}/T_s$ can be regarded as an integer due to the high resolution of channel taps in wide-band mmWave systems. For example, $T_s = 10\,\text{nm}$ when the bandwidth is 100 MHz, which means each tap can be discriminated well in the delay domain.

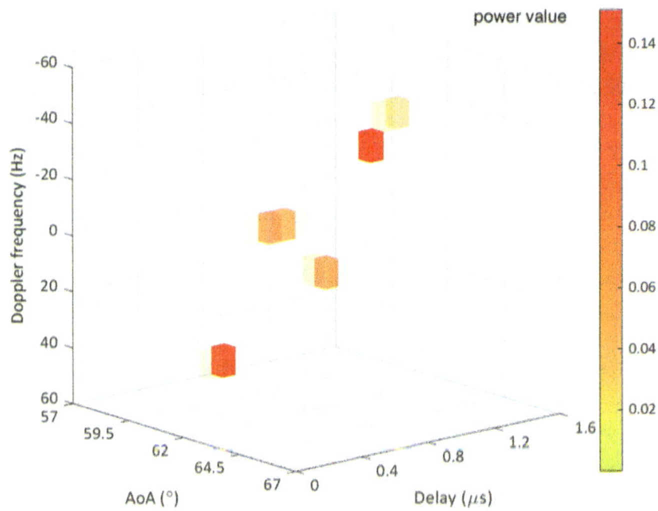

Fig. 7.5 Example of the 3-D channel power spectrum with $M = 256$, $N = 1024$, $T = 200$, and $|f_{u,l} \cdot f_{\text{ch}}| \leq 60\,\text{Hz}$

$$\lim_{M,T\to\infty} [\mathscr{P}_u]_{p,q,s} = \sum_{l=1}^{L} \rho_{u,l}\delta\left(p - \frac{Md_c \cos\theta_{u,l}}{\lambda} - \frac{M}{2}\right)\delta\left(q - \frac{\tau_{u,l}}{T_s}\right) \times \delta\left(s - f_{u,l}T - \frac{T}{2}\right). \tag{7.16}$$

An example of the ADD-PS obtained from above extraction method is shown in Fig. 7.5, where one colorful block represents a significant power value at the grid point of the corresponding AoA, delay, and Doppler frequency shift. Clearly, since each block contains both the propagation features and the activity information, the ADD-PS can be utilized as the 3-D fingerprint uniquely labelling the corresponding location and moving behavior.

In order to fulfill the objective of MIDAR, plenty of ADD-PSs corresponding to different locations with different behaviors are first collected and stored as reference points in a data set during the offline phase. Then, the detection is finished by comparing the acquired ADD-PS of the target terminal with the data set during the online phase. On the one hand, by transforming the problem into pattern recognition, the joint localization and behavior detection can be achieved via fingerprint matching methods. Since the ADD-PS is a multi-dimensional data array, to leverage its geometrical structure, the similarity criterion based on the chordal distance and Gaussian kernel as in [34] can be exploited. Compared with some other metrics simply used in vector spaces, this criterion depends on the linear subspaces coming from each unfolding matrix of a tensor and thus has potentials in providing superior performance. On the other hand, by regarding the problem as non-linear regression, the deep learning techniques such as CNN can be applied in MIDAR, which learns the complicated mapping from the fingerprint to the location as well as the behavior. The learning method is able to reduce the computational complexity during the online phase compared to the fingerprint matching approach, which needs to match the acquired fingerprint to every reference point in the data set.

In reality, the UL channel is acquired at the AP through the channel estimation while the estimation error due to the receive noise is inevitable. By employing the massive MIMO and carrying out the simple projection operation, not only the higher spatial resolution but also the better performance of power spectrum extraction can be obtained. Additionally, due to the significant attenuation of mmWave channels, the number of multipath components is limited, which means there are only a few dominant values in the ADD-PS and others are approximate to 0. This phenomenon can reduce the storage overhead of fingerprints and further eliminate the effect of channel estimation error by only saving those major elements in $\mathscr{P}_u$.

7.4 Active Approach

In this section, we will introduce active behavior recognition methods. In passive behavior recognition, the source data is the CSI which is primarily designed for the communication purpose. It cannot be arbitrarily configured due to the limitations in wireless communication protocols and Wi-Fi chips. In recent years, some researchers have tried to get out of the wireless CSI framework and accomplish user behavior recognition with specially designed radio frequency equipment. These works are thus called the active approaches. In the following subsections, we will introduce some recent works along this direction.

7.4.1 Millimeter Wave Radar Based Behavior Detection

In order to explore new human–computer interaction schemes, Google developed a new gesture sensing technology based on millimeter wave radar in 2015 [23]. They designed an all-in-one radar integrated circuit (IC) which integrated the functionalities of radar into one chip, as illustrated in Fig. 7.6, and designed a hardware abstract layer (HAL) to facilitate easy porting. This technology is called Soli.

Unlike the traditional radars which rely on spatially resolving multiple part of the target, Soli radar is mainly based on the high temporal resolution. In the following sections, we will introduce the Soli technology from Google in more detail.

Hand Model

Soli models the hand as a series of dynamic scattering centers as depicted in Fig. 7.7. This is a reasonable assumption as the wavelength is much smaller compared to the hand's spatial extent [35] and millimeter waves exhibit "optical" propagation characteristics. The hand reflects the millimeter waves and the angles of the reflected

Fig. 7.6 Soli: a millimeter wave radar system

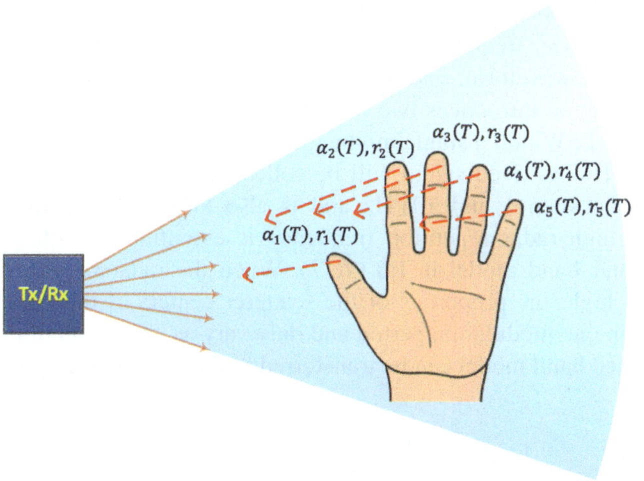

Fig. 7.7 The hand model used by Soli

waves are equal to the angles of the incident waves. This means that only parts of the reflected signals can be received. Therefore, the hardware can only observe parts of the hand and each part acts as a scatter at a different location. If the hand gesture changes, the received signal will also change correspondingly. A multipath channel impulse response (CIR) model can be utilized to characterize the channel between the transmitter and the receiver, which is mainly determined by the hand gesture. Further, we can mathematically describe the dynamic properties of the scatters at the hand as

$$d(r, T) = \sum_{j=1}^{N_{sc}} \alpha_i(T)\delta(r - r_i(T)), \tag{7.17}$$

where r denotes the distance between the transmitter/receiver and the scatter center, N_{sc} is the total number of observable scatter centers reflecting the transmitted millimeter waves back to the receiver, $r_i(T)$ is the distance between the i-th observable scatter center and the transmitter/receiver at time T, $\alpha_i(T)$ stands for the complex-valued reflectivity parameter, and $\delta(\cdot)$ is the Dirac delta function.

System Model

To realize a high temporal resolution, the Soli system sends out a periodically modulated waveform and measures the hand's responses at extremely high frame rates. Soli then extracts relevant features from the collected measurements and detects the complex and subtle hand motions. Specifically, the transmitted periodically modulated waveform is

$$x(t, T) = u(t - T)e^{j2\pi f_c t}, T = 0, T_0, 2T_0, \ldots, \tag{7.18}$$

where f_c is the carrier frequency, $u(t)$ denotes the complex envelope of one period of the modulated waveform, and T_0 stands for the signal repetition period. This transmission scheme introduces two time scales, i.e., the fast time scale t and the slow time scale T. Within a single modulation period T_0, the transmitted signal is a function of t. The received signal will be collected at a frequency of $F_r = 1/T_0$ which varies between 1 and 10 kHz and is also known as the radar repetition frequency. A high radar repetition frequency is essential for Soli to utilize the scattering center hand model in Eq. (7.17). When the radar repetition frequency is sufficiently high, the properties of the scatterer centers of the hand are almost constant within one modulation period and only vary as a function of the slow time T. The proposed hand model can be transferred to the following channel model:

$$h(t, T) = \sum_{i=1}^{N} \frac{\alpha_i}{r_i^4(T)} \delta\left(t - 2\frac{r_i(T)}{c}\right), \tag{7.19}$$

where the term $1/r_i^4(T)$ represents the path loss. The received signal is simply the superimposition of all the signals reflected by the scatterers, i.e.,

$$y_{raw}(t, T) = x(t, T) \otimes h(t, T) = \sum_{i=1}^{N_{sc}} y_{i,raw}(t, T), \tag{7.20}$$

where $\otimes$ denotes the convolution operation, $y_{i,raw}(t, T)$ denotes the signal component reflected by scatterer i and can be expressed as

$$y_{i,raw}(t, T) = \frac{\alpha_i(T)}{r_i^4(T)} u\left(t - 2\frac{r_i(T)}{c}\right) \exp\left(j2\pi f_c \left(t - 2\frac{r_i(T)}{c}\right)\right). \tag{7.21}$$

Signal Processing

Thanks to the fact that the signal is transmitted at two time scales, the received signal within one modulation period offers information about the ranges of the scatterer centers. Meanwhile, the variations at the slow time scale provide information about the dynamic properties of the scattering centers, such as velocity, change in geometry, etc. Accordingly, the processing of the received signal is also classed into two categories, i.e., fast time processing and slow time processing.

- *Fast Time Processing:* Soli mainly relies on the high temporal resolution introduced by the high radar repetition frequency. The processing method at the fast time scale is quite simple. Soli simply computes the fast time-frequency spectrogram decomposition of the raw receive signal as:

$$SP(t, f, T) = \int_{t}^{t+t_{win}} y_{raw}(\tau, T)e^{-j2\pi f\tau}d\tau, \tag{7.22}$$

  where t_{win} denotes the length of the time window. The range resolution of Soli with fast signal processing is limited. The detailed moving patterns of individual scattering centers of the hand are not resolvable at the fast time scale in most cases.
- *Slow Time Processing:* Since the transmitted signal is modulated, we need to carry out demodulation at the receiver side. Furthermore, modulation-specific filtering, such as the pulse compression and the matched-filtering, is also needed. After these preprocessing, the received signal can be expressed as

$$y_{rec}(t, T) = \sum y_{i,rec}(t, T), \tag{7.23}$$

  where $y_{i,rec}(t, T)$ denotes the response from the i-th scattering center and it is given by

$$y_{i,rec}(t, T) = \frac{\alpha_i(T)}{r_i^4(T)}\exp\left(j\frac{4\pi r_i(T)}{\lambda}\right)g\left(t - 2\frac{r_i(T)}{c}\right). \tag{7.24}$$

  where λ is the wavelength, $g(t)$ is the radar system point target response. Note $g(t)$ is determined by the modulation scheme, the transmission parameters, and the preprocessing steps. The shape of the radar system point target response function determines the range resolution. At the slow time scale, the dynamics of the scattering center cause the phase changes in the received signal $y_{rec}(t, T)$. These changes can be utilized to extract the responses from different scattering centers from the complex superposition as long as the radar repetition frequency is sufficiently high. Assuming the i-th scattering center moves from $r_i(T_1)$ to $r_i(T_2)$, the corresponding phase shift $\Delta\Phi_i(T_1, T_2)$ is given by

$$\Delta\Phi_i(T_1, T_2) = \frac{4\pi}{\lambda}(r_i(T_2) - r_i(T_1)) \mod 2\pi. \tag{7.25}$$

  When the radar repetition frequency is very high, the time interval between two adjacent frames is small. If the time interval is less than the coherent time T_{cpi}, the velocity in this time interval can be regarded constant, i.e.,

$$\frac{dr_i(T)}{dT} = v_i(T) \approx v_i, \text{ for } T_{cpi} > T_0. \tag{7.26}$$

The phase changes lead to the following Doppler frequency

$$f_{D,i}(T) = \frac{1}{2\pi}\frac{d\Phi_i(T)}{dT} = \frac{2v_i(T)}{\lambda}. \tag{7.27}$$

As different scattering centers exhibit different moving speeds, the Doppler frequencies of multiple scattering centers can be resolved by computing the spectrum of $y_{rec}(t, T)$ within each fast time bin over the coherent processing window T_{cpi}, i.e.,

$$S(t, f, T) = \int_T^{T+T_{cpi}} y_{rec}(t, \tau)e^{-j2\pi f\tau}d\tau. \tag{7.28}$$

Combining Eqs. (7.27) and (7.28), the fast time-frequency mapping $S(t, f, T)$ can be transformed into the range-Doppler mapping as follows:

$$RD(r, v, T) = S\left(\frac{2r}{c}, \frac{2v}{\lambda}, T\right). \tag{7.29}$$

$RD(r, v, T)$ is a three-dimensional array which maps the reflected energy from each scattering center to its range r and velocity v at time T. Thus two scattering centers are distinguishable in the range-Doppler array as long as one of the following conditions can be met:

- The range difference is greater than the range resolution, which is determined by the radar spatial resolution, $res_r = c/(2\mathtt{BW})$, where c is the speed of light and BW is the bandwidth of the transmitted waveform;
- The separation in velocity is greater than the Doppler velocity resolution which is given by $\lambda/(2T_{cpi})$;
- They are detectable only in different coherent processing time windows.

Soli does not excel in spatial resolution as it does not adopt a large bandwidth. Instead, Soli relies on the high temporal resolution to bring in high Doppler velocity resolution.

Feature Extraction and Gesture Recognition

After processing the raw data as described above, Soli extracts miscellaneous features and puts them into machine learning algorithm to perform gesture recognition. The features Soli extracted can be divided into three groups:

- *Explicit Scattering Center Tracking Features*: Soli estimates the number of resolvable scattering centers $\bar{N}_{sc}$, and for each scattering center, estimates its range $r_i(T)$, velocity $v_i(T)$, acceleration $dv_i(T)/dT$, and the reflectivity $\alpha_i(T)$;
- *Low Level Descriptors of the Physical RF Measurement*: Simple velocity features cannot make the scattering centers distinguishable as the fingers usually moves

at similar speed. Thus Soli extracts low level abstracts features which depict the energy distribution in the signal transformation space to describe the finger dynamics. Those features includes: velocity profile centroid, relative displacement, velocity profile dispersion, range profile dispersion, total instantaneous energy, and total time-varying energy;

- *Data-Centric Machine Learning Features*: In order to reduce the dimensionality of the data, Soli defines a region of interest (ROI) in transformation, and then extracts data-centric features within the ROI, such as range-Doppler multichannel integration, range-Doppler multichannel derivative, range-Doppler temporal derivative, spectrogram multichannel integration, and other features.

Since Soli aims to recognize dynamic hand gestures instead of static hand postures, features from multiple frames are needed. Soli extracts some features in a sliding temporal window and stores them in a temporal buffer. Besides, meta features such as means, root mean square deviation, and other basic static can be computed based on the current contents of the buffer. The extracted features are then input into a machine learning engine to recognize the hand gesture. Soli further introduces a filter after the machine learning algorithm. Specifically, Soli first uses a machine learning algorithm to infer the gesture coarsely and then utilizes a Bayesian filter to improve the accuracy further.

Different machine learning algorithms such as SVM, HMM, or CNN, can be used. When the computing power is limited, a random forest classifier can also be chosen. The Bayesian filter predicts gesture $\hat{g}$ among a series gesture g_k, $1 \leq k \leq K$, according to the following rule:

$$\hat{g} = \arg\max P(g_k | \boldsymbol{f}), \tag{7.30}$$

where $P(g_k|\boldsymbol{f})$ is the posterior probability for gesture k given the feature vector $\boldsymbol{f}$. In particular, it can be computed as

$$P_{g_k|f} = \frac{P(\boldsymbol{f}|g_k)P(g_k)}{\sum_i P(\boldsymbol{f}|g_i)P(g_i)}, \tag{7.31}$$

where $P(\boldsymbol{f}|g_k)$ is the raw prediction likelihood output by classifier and $P(g_k)$ is the prior probability of g_k. The value of $P(g_k)$ at time T can be calculated by

$$P(g_k^{(T)}) = z_k^{(T)} \sum_{n=1}^{N} w_n P(\boldsymbol{f}^{(T-n)} | g_k^{(T-n)})^{(T-n)}, \tag{7.32}$$

where $z_k^{(T)}$ is the contextual prior specified by high level application and w_n is the filter weight for previous prediction.

A human commuter interference with the Soli chip was demonstrated in the work [23]. When a user performs a gesture, the user imagines he is operating a virtual tool, such as a button or a slider. The random forest was utilized to recognize the gesture in real time. The reported per-gesture recognition accuracy is around 92%.

7.4.2 Acoustics Based Localization and Motion Tracking

In this subsection, we will introduce the phase-based acoustic motion tracking (PAMT) system. In PAMT, the sources are static speakers, which transmit inaudible acoustic signals. The receiver is the microphone of a cell phone, which collects the acoustic signal frame by frame and the motion of the cellphone can be tracked based on the phase changes of the acoustic signals. PAMT can achieve millimeter-level accuracy for localization and motion tracking. The measurement errors are less than 2 mm in one-dimensional scenarios and 4 mm in two-dimensional scenarios.

To achieve high accuracy motion tracking, two challenges have to be addressed. One is to calibrate the phase offsets caused by the asynchronous system clocks of the transmitters and the receiver. The other one is to deal with the multiple paths since the received signals are a superposition of the desired LOS signals and the NLOS interference signals.

Calculating Phase Changes

Assume that a single tone acoustic signal, for instance, $A\cos(2\pi f_c t)$, is transmitted from the transmitter. The mobile device acquires the acoustic signal $R_i(t)$ via its microphone. The signal first goes through a band-pass filter (BPF) with a center frequency of f_c and a narrow pass band. The filtered signal $R_\alpha(t)$ is then delayed by a quarter of fundamental-wave period. The delayed signal is denoted by $R_\beta(t)$. See also Fig. 7.8 and $R_\alpha(t)$, $R_\beta(t)$ can be written as

$$\begin{aligned} R_\alpha(t) &= A'\cos(2\pi f_c t - \tau) \\ R_\beta(t) &= A'\sin(2\pi f_c t - \tau) \end{aligned}. \tag{7.33}$$

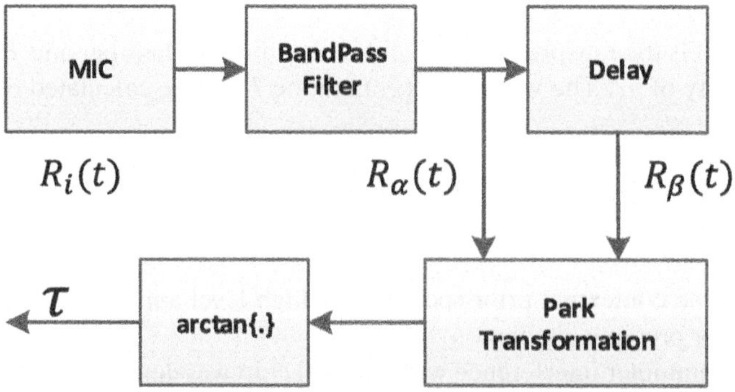

Fig. 7.8 Phase changes calculation

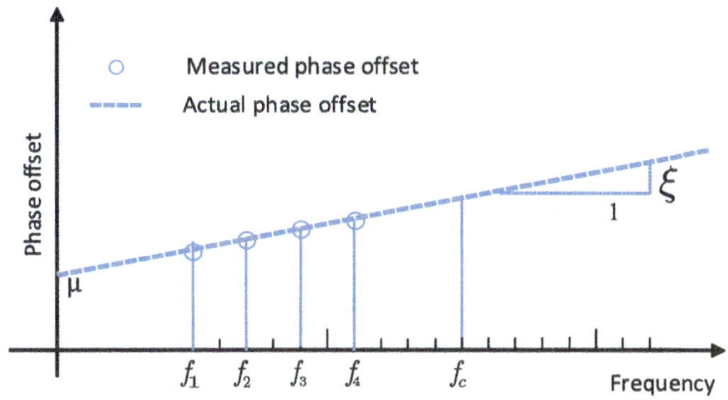

Fig. 7.9 The phase offset in frequency domain caused by the asynchronous system clocks

PAMT system then uses Park transformation to calculate the phase changes, specifically, multiply a transformation matrix to $R_\alpha(t)$ and $R_\beta(t)$, i.e.

$$\begin{bmatrix} R_d(t) \\ R_q(t) \end{bmatrix} = \begin{bmatrix} \cos(2\pi f_c t) & \sin(2\pi f_c t) \\ -\sin(2\pi f_c t) & \cos(2\pi f_c t) \end{bmatrix} \begin{bmatrix} R_\alpha(t) \\ R_\beta(t) \end{bmatrix} = \begin{bmatrix} A' \cos(\tau) \\ A' \sin(\tau) \end{bmatrix}. \tag{7.34}$$

The phase delay at time t can be calculated by $\tau = \arctan(R_q(t)/R_d(t))$.

However, the asynchronous clocks will cause phase offset in the received signal. According to the experiments, the phase offset is related to the frequency, as shown in Fig. 7.9. Therefore, we measure the phase offsets at several frequencies from a source at first, and then use a linear fitting to estimate the slop and intercept of the linear phase shift in frequency domain. If the multiple speakers share the same system clock, we can estimate the slop and the intercept via the signals from one of the speakers, and apply them to the signals of all the speakers. After that, we can compensate the frequency offset caused by the asynchronous clocks as follows:

$$\tau^{\texttt{adjusted}} = \tau - wt, \tag{7.35}$$

where τ is the raw phase change at frequency f_c, $w = \xi f_c + \mu$ is the phase offset at frequency f_c, ξ and μ are the estimated slop and intercept, respectively.

Combating Multipath Effect

In practical systems, the received signals are a superposition of the LOS signals and the NLOS signals. This multipath effect would lead to periodic attenuation of the received signals, which brings error in estimating the moving distance and the direction.

To combat multipath effect, we leverage the fact that the multiple paths do not always affect the phase- based measurements of different frequencies severely at the same time due to their different wavelengths and phases. Therefore, we let the speakers send signals at different frequencies, and choose the signal which is less affected by the multipath effect in each frame.

To measure the impacts of multipath effect, we define a multipath effect ration (MER). If we plot the normalized signal's trace diagram in a frame using $R_\alpha(t)$ and $R_\beta(t)$. We find that

- When the mobile device is static, the trace of the signal at each frequency is a circle.
- When the device moves, the trace of the signal at each frequency is a circular ring. The width of the ring indicates the impact of multipath effects.

This is because the radius of the trace is a function of the amplitude and phase delay of each path. When the device is static, these parameters remain constant, and so will the radius, which results in a circle. When the devices moves, the radius will change over the time, which results in a ring for each frequency. The width of the ring reflects the dynamics in the radius, which is decided by the multipath effects. Therefore, the MER of the signal at a specific frequency is defined as the ratio of mean inter radius to the mean radius of the ring. Larger MER means less impact of the multipath effect. Thus the signal with the largest MER is chosen to estimate the moving distance in each frame.

One example is illustrated in Fig. 7.10. The speaker continuously transmits acoustic signals at 17.2 and 18.8 kHz simultaneously. The MER over time of the two signals are plotted in Fig. 7.10a, and the normalized signal's trace of the chosen frames (Frame A and Frame B indicated in Fig. 7.10a) are shown in Fig. 7.10c, d respectively. The MER of the signal at 17.2 kHz is larger in Frame A while is smaller in Frame B. Therefore, the signal at 17.2 kHz is chosen to estimate the distance in Frame A while the signal at 18.8 kHz is chosen in Frame B. The estimated distance is depicted in Fig. 7.10b. The four lines indicate the results of only using the signal at 17.2 kHz, only using the signal at 18.8 kHz, using the MER to choose the used signal, and the ground truth, respectively. The results of using the MER to choose the used are close to the ground truth, which prove the effectiveness of this multipath combining method.

Robust Phase-Based Ranging

In each frame, the moving distance Δd can be calculated according to the phase change of the chosen signal with the largest MER.

$$\Delta d = -\frac{\Delta\theta + 2\pi n}{2\pi}\lambda, \tag{7.36}$$

where $\Delta\theta$ is the wrapped phase change relative to initial phase, n is an integer accounting for the phase wrapping, n increases (decreases) by one if the phase varies from π to $-\pi$ (from $-\pi$ to π), λ is the wavelength.

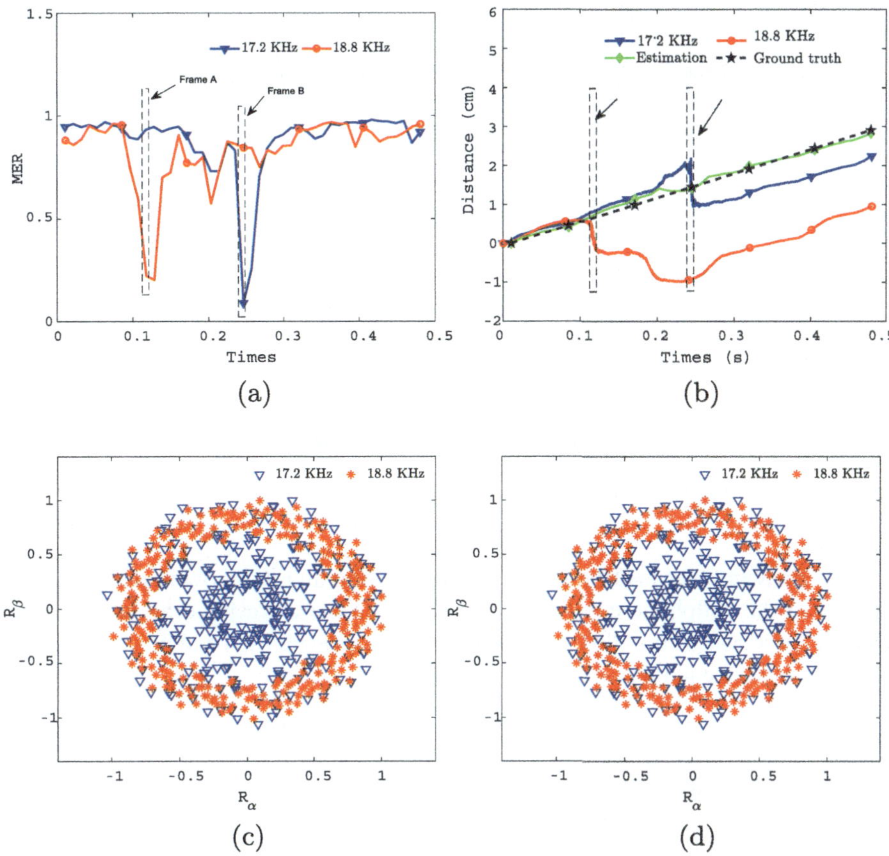

Fig. 7.10 An example of combating multipath effect. (**a**) The MER of the two signals. (**b**) The tracking results. (**c**) The traces of Frame A. (**d**) The traces of Frame B

Knowing the moving distance is not enough to track the cellphone because the initial position is unknown. PAMT proposed a new method to estimate the position of the reference point and take it as the initial position. Assume that two speakers locate at point A and point B respectively. The cellphone moves parallel to the line AB. Point C and point D are the chosen reference points at which the distance to one of the speakers is minimum during the moving. Notation a, b, and c denotes the distance from point A and to C, point A to point B, point A to point D, respectively. The difference between a and c is denoted by $d_{ac} = c - a$, which can be measured when the cellphone moves from point C to point D. Then $a = \frac{b^2 - d_{ac}^2}{2d_{ac}}$. Given the distance and the direction of the two speakers, the cellphone can find the location of the reference points without additional measurements (see also Fig. 7.11).

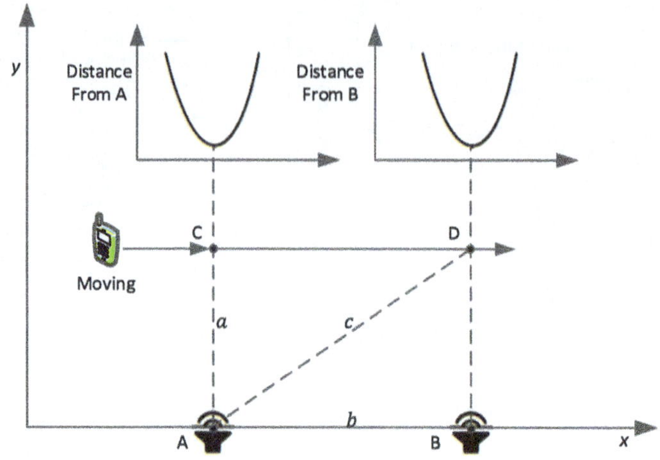

Fig. 7.11 Estimating the location of the reference points

After knowing the position of the reference point and the moving distance, the position of the cellphone can be derived based on the corresponding geometrical relationship. Considering the estimation error, least square (LS) can be used here to avoid solving the equation set.

7.5 Conclusion

This chapter summarizes current user behavior recognition methods which can be used in fog-enabled smart home scenario. The passive approach which uses current commercial Wi-Fi devices is easy to implement but the recognition capability is limited. To achieve higher recognize accuracy with radio, millimeter wave must be used. This will not be an issue in the future as millimeter wave communication is an inevitable trend in the future.

In the future, behavior recognition and fog-enabled smart home are sure to be integrated more closely. Most of the existing works focus only on situations with single transmitter and relatively small numbers of receivers. The fact is that, however, the reflected signal can be received in varieties of directions in a smart house full of IoT nodes. The accuracy and the range of detection can be further improved with these signals combined. Moreover, most, if not all, of the current work employ center processing as far as we have known. Distributed computation in fog-enabled smart can also be a boost in facilitating behavior recognition considering the fact that, each IoT node has its own observation and center processing increases extra communication load, thus resulting in the waste of computation resources. Every IoT node equipped with certain computing capability can process raw data, transfer compressed features to center processing unit, or even make local decisions. This could be a promising technique to improve behavior recognition in the future work.

References

1. Ma J, Wang H, Zhang D, Wang Y, Wang Y (2016) A survey on wi-fi based contactless activity recognition. In: 2016 International IEEE conferences on ubiquitous intelligence computing, advanced and trusted computing, scalable computing and communications, cloud and big data computing, internet of people, and smart world congress (UIC/ATC/ScalCom/CBDCom/IoP/SmartWorld). pp 1086–1091
2. Popoola OP, Wang K (2012) Video-based abnormal human behavior recognition—a review. IEEE Trans Syst Man Cybern Part C Appl Rev 42(6):865–878
3. Gavrilova ML, Wang Y, Ahmed F, Polash Paul P (2018) Kinect sensor gesture and activity recognition: new applications for consumer cognitive systems. IEEE Consum Electron Mag 7(1):88–94
4. Karantonis DM, Narayanan MR, Mathie M, Lovell NH, Celler BG (2006) Implementation of a real-time human movement classifier using a triaxial accelerometer for ambulatory monitoring. IEEE Trans Inf Technol Biomed 10(1):156–167
5. Lara OD, Labrador MA (2013) A survey on human activity recognition using wearable sensors. IEEE Commun Surv Tutorials 15(3):1192–1209
6. Hegde N, Bries M, Swibas T, Melanson E, Sazonov E (2018) Automatic recognition of activities of daily living utilizing insole-based and wrist-worn wearable sensors. IEEE J Biomed Health Inform 22(4):979–988
7. Wang Y, Liu J, Chen Y, Gruteser M, Yang J, Liu H (2014) E-eyes: device-free location-oriented activity identification using fine-grained wifi signatures. In: Proceedings of the 20th annual international conference on mobile computing and networking, MobiCom'14. ACM, New York, pp 617–628
8. Wang W, Liu AX, Shahzad M, Ling K, Lu S (2015) Understanding and modeling of WiFi signal based human activity recognition. In: Proceedings of the 21st annual international conference on mobile computing and networking, MobiCom'15. ACM, New York, pp 65–76
9. Liu X, Cao J, Tang S, Wen J (2014) Wi-sleep: contactless sleep monitoring via WiFi signals. In: 2014 IEEE Real-time systems symposium, pp 346–355
10. Liu X, Cao J, Tang S, Wen J, Guo P (2016) Contactless respiration monitoring via off-the-shelf WiFi devices. IEEE Trans Mob Comput 15(10):2466–2479
11. Liu J, Wang Y, Chen Y, Yang J, Chen X, Cheng J (2015) Tracking vital signs during sleep leveraging off-the-shelf WiFi. In: Proceedings of the 16th ACM international symposium on mobile Ad Hoc networking and computing, MobiHoc'15. ACM, New York, pp 267–276
12. Khan UM, Kabir Z, Hassan SA, Ahmed SH (2017) A deep learning framework using passive WiFi sensing for respiration monitoring. In: GLOBECOM 2017–2017 IEEE global communications conference, pp 1–6
13. Wang H, Zhang D, Wang Y, Ma J, Wang Y, Li S (2017) Rt-fall: a real-time and contactless fall detection system with commodity WiFi devices. IEEE Trans Mob Comput 16(2):511–526
14. Wang Y, Wu K, Ni LM (2017) Wifall: device-free fall detection by wireless networks. IEEE Trans Mob Comput 16(2):581–594
15. Zheng X, Wang J, Shangguan L, Zhou Z, Liu Y (2016) Smokey: ubiquitous smoking detection with commercial WiFi infrastructures. In: IEEE INFOCOM 2016—the 35th annual IEEE international conference on computer communications, pp 1–9

16. Zheng X, Wang J, Shangguan L, Zhou Z, Liu Y (2017) Design and implementation of a CSI-based ubiquitous smoking detection system. IEEE/ACM Trans Netw 25(6):3781–3793
17. Abdelnasser H, Youssef M, Harras KA (2015) Wigest: a ubiquitous WiFi-based gesture recognition system. In: 2015 IEEE conference on computer communications (INFOCOM), pp 1472–1480
18. Tan S, Yang J (2016) Wifinger: leveraging commodity WiFi for fine-grained finger gesture recognition. In: Proceedings of the 17th ACM international symposium on mobile Ad Hoc networking and computing, MobiHoc'16. ACM, New York, pp 201–210
19. Ali K, Liu AX, Wang W, Shahzad M (2017) Recognizing keystrokes using WiFi devices. IEEE J Sel Areas Commun 35(5):1175–1190
20. Qian K, Wu C, Zhou Z, Zheng Y, Yang Z, Liu Y (2017) Inferring motion direction using commodity wi-fi for interactive exergames. In: Proceedings of the 2017 CHI conference on human factors in computing systems, CHI'17. ACM, New York, pp 1961–1972
21. Qian K, Wu C, Yang Z, Liu Y, Jamieson K (2017) Widar: decimeter-level passive tracking via velocity monitoring with commodity wi-fi. In: Proceedings of the 18th ACM international symposium on mobile Ad Hoc networking and computing, Mobihoc'17. ACM, New York, pp 6:1–6:10
22. Qian K, Wu C, Zhang Y, Zhang G, Yang Z, Liu Y (2018) Widar2.0: passive human tracking with a single wi-fi link. In: Proceedings of the 16th annual international conference on mobile systems, applications, and services, MobiSys'18. ACM, New York, pp 350–361
23. Lien J, Gillian N, Emre Karagozler M, Amihood P, Schwesig C, Olson E, Raja H, Poupyrev I (2016) Soli: ubiquitous gesture sensing with millimeter wave radar. ACM Trans Graph 35(4):142:1–142:19
24. Wang S, Song J, Lien J, Poupyrev I, Hilliges O (2016) Interacting with soli: exploring fine-grained dynamic gesture recognition in the radio-frequency spectrum. In: Proceedings of the 29th annual symposium on user interface software and technology, UIST'16. ACM, New York, pp 851–860
25. Wei T, Zhang X (2015) mTrack: high-precision passive tracking using millimeter wave radios. In: Proceedings of the 21st annual international conference on mobile computing and networking, MobiCom'15. ACM, New York, pp 117–129
26. Nandakumar R, Iyer V, Tan D, Gollakota S (2016) Fingerio: using active sonar for fine-grained finger tracking. In: Proceedings of the 2016 CHI conference on human factors in computing systems, CHI'16. ACM, New York, pp 1515–1525
27. Wang W, Liu AX, Sun K (2016) Device-free gesture tracking using acoustic signals. In: Proceedings of the 22Nd annual international conference on mobile computing and networking, MobiCom'16. ACM, New York, pp 82–94
28. Halperin D, Hu W, Sheth A, Wetherall D (2011) Tool release: gathering 802.11n traces with channel state information. SIGCOMM Comput Commun Rev 41(1):53–53
29. del Peral-Rosado JA, Raulefs R, López-Salcedo JA, Seco-Granados G (2018) Survey of cellular mobile radio localization methods: from 1G to 5G. IEEE Commun Surv Tutorials 20(2):1124–1148. Secondquarter
30. Wei Z, Zhao Y, Liu X, Feng Z (2017) DoA-LF: a location fingerprint positioning algorithm with millimeter-wave. IEEE Access 5:22678–22688
31. Lin Z, Lv T, Mathiopoulos PT (2018) 3-d indoor positioning for millimeter-wave massive MIMO systems. IEEE Trans Commun 66(6):2472–2486
32. Shahmansoori A, Garcia GE, Destino G, Seco-Granados G, Wymeersch H (2018) Position and orientation estimation through millimeter-wave MIMO in 5G systems. IEEE Trans Wirel Commun 17(3):1822–1835
33. Abu-Shaban Z, Zhou X, Abhayapala T, Seco-Granados G, Wymeersch H (2018) Error bounds for uplink and downlink 3D localization in 5G millimeter wave systems. IEEE Trans Wirel Commun 17(8):4939–4954
34. Cyganek B, Krawczyk B, Wozniak M (2015) Multidimensional data classification with chordal distance based kernel and support vector machines. Eng Appl Artif Intell 46(PA):10–22
35. Keller JB (1962) Geometrical theory of diffraction. J Opt Soc Am 52(2):116–130

Chapter 8
Conclusions

A huge variety of IoT devices have been widely deployed in our modern society for environmental monitoring, infrastructure management, intelligent manufacturing, operation optimization, safety and surveillance, remote healthcare, and so on, which continue to generate more and more data and, therefore, demand efficient resource sharing, data processing, information extraction, and decision making in real time. This problem becomes much harder in mobile environments when tens of billions of smartphones and vehicles are connecting to communication networks for different mobile applications and interactive services, such as online gaming, high-resolution video streaming, augmented/virtual reality, and autonomous driving.

Despite different industrial sectors and service scenarios, it is a clear trend that IoT technologies, applications, and business models are fast evolving towards more intelligent and sophisticated tasks such as cross-domain data analysis, pattern recognition, and behavior prediction, in addition to traditional simpler tasks such as data sensing, collection, and representation. In order to promote this trend, we should develop a user-centric approach to enable different IoT applications being automatically configured and customized according to specific user preferences, service scenarios, and performance requirements in various industrial domains. This approach requires ubiquitous deployments of multi-tier computing resources and adaptive algorithms at global, regional, local, and device levels. However, as a centralized solution, cloud computing alone cannot effectively support intelligent IoT applications and services in different domains and scenarios, due to limited communication bandwidth, intermittent network connectivity, and strict delay constraints.

As an extension of cloud computing, fog computing is recently proposed and deployed along the continuum from the cloud to things [1, 2], thus enabling integration and collaboration of different computing resources at the cloud, in the network, at the edge, and on the things (devices). In other words, fog computing provides a new architecture to effectively pool dispersive computing resources at different levels, and then bridge sophisticated service requirements with best

Y. Yang et al., *Fog-Enabled Intelligent IoT Systems*,
https://doi.org/10.1007/978-3-030-23185-9_8

available resources in the neighborhood. By doing this, it ensures timely data processing, situation analysis, and decision making at the locations very close to where the data is generated and should be used. Some aforementioned infrastructure challenges, such as network connectivity, communication bandwidth, and service latency, can be successfully addressed by fog computing for supporting more intelligent IoT applications and services.

As discussed in previous IoT applications, fog computing manages and utilizes the multi-tier computing, communication, and storage resources within the networks between the cloud and things. It enables more intelligent services and makes those applications more accessible, flexible, efficient, and cost-effective. Despite these advantages, we are still facing many technical challenges in realizing the vision and the full potential of fog computing. Here are some key research topics deserving a further in-depth study.

1. Service Architecture and Orchestration: How to design a comprehensive architecture with new business models to consolidate the ecosystem of shared resources and collaborative services? How to timely identify them and orchestrate a sophisticated service to meet dynamic user requirements?
2. Interoperability and Interface: How to ensure smooth interoperations in vertical and horizontal directions, i.e., between cloud, fog, and edge, and between heterogeneous nodes? How to define proper interfaces between different network entities and implement unified cross-domain policies and protocols?
3. Security and Privacy: How to safely find, use, and trust anonymous computing nodes in your neighborhood? How to enjoy the benefits and convenience of using shared computing resources without compromising any userąŕs identity or personal data?
4. Quality of Service: How to guarantee service quality in mobile environments and multi-user scenarios, where dynamic network topologies, frequent node handovers, and competing service requests will degrade system performance and user experience.
5. Resource Virtualization and Management: How to accurately characterize the capabilities and dynamics of different network resources in real time? How to effectively model, virtualize, and manage them at device, local, regional, and global levels?
6. Task Scheduling and Performance Optimization: How to divide a complex task or service into several simpler ones, which could be processed in more efficient ways by nearby computing resources? How to match sub-tasks with feasible nodes for achieving the optimal performance and economic targets?

References

1. Chiang M, Zhang T (2016) Fog and IoT: an overview of research opportunities. IEEE Internet Things J 3(6):854–864
2. Yang Y (2019) Multi-tier computing networks for intelligent IoT. Nat Electron 2(1):4

Index

A
Acoustics based localization
  multipath effect, 205–206
  phase changes, 204–205
  robust phase-based ranging, 206–208
Active approach
  acoustics based localization (*see* Acoustics based localization)
  millimeter wave radar based behavior detection, 198–203
Artificial intelligence (AI), 3, 8, 36, 42, 45, 46, 51, 59, 105
Autonomous driving (AD), 56, 57, 172, 180–181

B
Behavior recognition, *see* User behavior recognition

C
Centralized SON (C-SON), 135–136
Collaborative robot system, 52–53
Communication network, 170–172
Computation offloading, 104, 150–151
  algorithm, 77
  joint optimization, 151–158
  SLAM, 106
  TNs, 94
Convergence, 13, 95
Cooperative driving, 175, 181
Coverage and capacity optimization (CCO)
  deep-learning, 147–148
  fog-computing enabled RAN, 147
  self-optimization (*see* Self-optimization)

D
Distributed SON (D-SON), 136–137
Duty cycle, 16

E
Environmental monitoring, 23–24, 39, 110, 113

F
Fog computing
  architecture
    multi-tier computing network, 45–48
    reference, 42–45
  benefits, 61
  challenges, 39–40
  cloud-based computing model, 173
  cloud-to-thing continuum, 40
  definition, 40
  edge devices, 144
  MEC, 41
  mobile network, 41
  on-demand mobility management, 139
  OpenFog, 42
  resource-limited devices, 39
  SDN and NFV, 134
  SDWN, 136

Y. Yang et al., *Fog-Enabled Intelligent IoT Systems*,
https://doi.org/10.1007/978-3-030-23185-9

Fog computing (*cont.*)
 self-optimization (*see* Self-optimization)
 shared and flexible communication, 61
 signaling, 142
 technologies
  AI, 51–52
  computing, 49
  networking, 48
  storage, 49–50
  virtualization, 50–51
 VMs, 41
Fog-enabled solution
 architecture, 173–177
 communication networks, 178–180
 distributed resources, 178–180
 key technologies, 173–177
 virtualized and intensive vehicle stations, 177–178
Fog networks
 computation tasks, 150
 delay-sensitive applications, 61
 heterogeneous, 61
 MTMH, 65–68
 non-splittable tasks, 64
 robot SLAM, 53
 wireless access, 102
Fog nodes (FNs)
 computation offloading (*see* Computation offloading)
 definition, 40
 features and functions, 174–177
 geographical distribution, 134
 hierarchical control system, 51
 IoT devices, 48
 resource pool, 61
 RSUs, 180
 sensors, 48
Formation control, 125, 126, 128
FSDN VANET, 178, 179

G
Game formulation
 paired offloading of multiple tasks, 75–78
 parallel offloading of splittable tasks, 79–87
Generalized Nash equilibrium problem (GNEP), 64, 73, 79, 85, 96

H
Handover optimization, 142–143
Hybrid SON (H-SON), 137–138

I
Intelligent transportation system (ITS)
 applications, 32
 architecture, 165–167
 communication network, 170–172
 countries, 163–165
 data sharing, 57
 deployment environment, 57–58
 explosive computing workload, 56
 fog-enabled, 65
 functions, 31–32
 key components, 165–167
 on-vehicle computing, 168–170
 requirements, 167–168
 scenario, 31
 security and privacy, 56–57
 traffic congestion, 163
 vehicle categories, 167–168
Inter-cell interference coordination (ICIC)
 conventional single-tier cellular networks, 139
 eICIC, 144
 in RANs, 144
 self-coordination (*see* Self-coordination)
 user scheduling, 146
Interference-aware radio resource allocation, 144–146
Internet of things (IoT)
 AI technologies, 3
 applications, 2
  infrastructure health monitoring, 25, 26
  public safety surveillance, 26–28
  smart city, 28–29
  smart manufacturing, 29–30
 business models, 1
 challenges
  big data, 33
  centralized processing, 34
  energy requirement, 33
  operation costs, 34
  privacy, 35
  robustness, 35–36
  scalability and diversity, 33
  security, 34–35
 cloud computing, 211
 communication resources, 1
 comparisons, 6
 connectivity for anything, 1
 data collection and analysis, 2
 fast-growing demands, 3
 industrial sectors, 211
 required data rate *vs.* range capacity, 4
 research topics, 212

J
Joint optimization of computation offloading
  cloud processing, 154–158
  FN, 152
  fog-computing enabled wireless networks, 151
  mobile applications, 152
  processing
    fog, 153–154
    local, 153
  remote processing, 152

L
Load balancing, 149
Long range radio (LoRa)
  bandwidth, 14
  communication protocol, 13
  IoT application, 2
  LoRaWAN network, 16
  municipal operations, 15
  network architecture, 15
  server architecture, 14
  unlicensed LPWAN, 20

M
Massive MIMO based detection and ranging (MIDAR), 194–198
Matching theory, 73–75
Millimeter wave radar based behavior detection
  feature extraction, 202–203
  gesture recognition, 202–203
  hand model, 198–200
  signal processing, 200–202
  system model, 200
Mobile/multi-access edge computing (MEC), 41
Mobility
  control, 46
  handover optimization, 142–143
  management and procedures, 139–142
  robustness, 142–143
  self-optimization (*see* Self-optimization)
Mobility load balancing (MLB), 138–139
Multi-robot smart factory
  emergency management, 119–125
  system architecture
    cloud network layer, 112–113
    fog network layer, 112
    physical layer, 111–112
    supervisory control layer, 113
  warehouse management (*see* Warehouse management)
Multi-robot system
  AGVs, 99
  cloud-to-thing continuum, 99
  fleet formation
    fog-enabled solution, 127–130
    system architecture, 125–127
Multi-task multi-helper (MTMH)
  energy-minimization workload, 63
  FNs, 61
  fog networks, 65–68
  GNEP, 73
  groups, 64
  heterogeneous fog network, 62
  HNs and TNs, 61, 62
  large-scale fog networks, 64
  NOMA, 63
  non-splittable task, 68–69
  offloading
    paired, 68–69
    parallel, 69–70
  potential game, 70–72
  splittable task, 69–70
Multi-tier computing network, 42, 45–48, 59, 211

N
Narrowband IoT (NB-IoT)
  FOTA, 21
  in 4G, 2, 18–21
  global IoT market, 20
  IoT applications, 20
  LPWAN technology, 18
  network architecture, 19
  objectives, 19
  pre-specified factor, 5
  SCEF, 19
National ITS Master Plan, 165
Near field communication (NFC), 2, 4–6, 9–11, 48
Network edge caching, 149–150
Network resources
  computation offloading, 150–151
  edge caching, 149–150
  fog-computing enabled RANs, 149
  joint optimization of computation, 151–158
  proliferation and developments, 148
Non-orthogonal multiple access (NOMA), 63

O
On-vehicle computing, 168–170
OpenFog reference architecture, 42–45, 59
Operational silos, 57

P
Paired Offloading of Multiple Tasks (POMT), 64, 75–78, 89–96
Passive approach
MIDAR, 194–198
signal detection, 189–192
wi-fi based solution, 192–194
Pervasive computing, 2, 35
Phase-based acoustic motion tracking (PAMT), 204, 205, 207
Price of anarchy (PoA), 87–89, 91
Public safety surveillance, 26–28, 32

R
Radio frequency identification (RFID)
adoption, 8
advantages, 7
architecture, 7
components, 7
data acquisition, 2
IoT data collection, 32
reliability, 5
researchers and scientists, 9
RFID-based anti-theft vehicle anti-theft device, 8
standardization, 5
Resource allocation
edge cloud based robotic workflow, 121
homogeneous and heterogeneous fog networks, 63
interference-aware radio, 144–146
joint optimization, 151–158
radio, 63
RAT, 54
user scheduling, 144

S
Self-coordination
fog-computing enabled RANs, 144
interference-aware radio resource allocation, 144–146
multi-tier HetNet, 143
Self-optimization
CCO (*see* Coverage and capacity optimization)
MLB, 138–139
mobility management and procedures, 139–142
network resources (*see* Network resources)
Self-organizing networks (SONs)
C-SON, 135–136
design, 135
D-SON, 136–137
functionalities, 54, 55
H-SON, 137–138
multi-vendor and multiple RAT, 134
3GPP Release 8, 133
Shared vehicle, 181–183
Sigfox
comparison of IoT connecting technologies, 6
French global network operator, 16
IoT applications, 17
network architecture, 17
pallets tracking, 18
price-competitive connectivity, 18
sub-GHz spectrum, 16
Signal detection, 189–192
Simultaneous location and mapping (SLAM)
environment and estimation, 52
fog-enabled solution, 53, 102–109, 125
function view, 106
maps, 52
step modules, 106
system architecture, 100–101
technical challenges, 101–102
Smart city, 1, 15, 23, 28–29, 47
Smart highways, 57, 173, 175
Smart home, 58–59
fog-enabled intelligent IoT system, 185
human behavior recognition, 185
radio-based behavior recognition, 185
researches, 185
user behavior recognition, 187–189
Smart sensing, 2
Soli, 186, 188, 198–203
Southbound interface, 103, 105
Strategy profile, 70, 71
System average delay, 88–93

T
Task nodes (TNs)
beneficial, 93–95
broadcasting parameters, 77, 86
computation offloading, 94
cost function, 75
dynamic programming, 64
heterogeneous sensors, 63
pairing strategies, 69
process tasks, 61
system average delay, 91
Time sensitive networking (TSN), 48

U
User behavior recognition
  fog-enabled smart home, 208
  human, 185
  smart home, 187–189
  Wi-Fi CSI framework, 189

V
Virtual machines (VMs)
  cloud, 118
  isolation, 50
  proxy, 143
  shared power, 112
  virtualization, 41

W
Warehouse management
  automated parcel, 115–116
  automation support, 113
  energy consumption, 117
  experimental results, 118–119
  joint collaboration, 117
  parcel sorting and distribution application, 114
  sensors, 113
  shared architecture work, 115
Web of things (WoT), 5, 21–23
Wi-fi based solution, 192–194
Wireless communication network, 24, 54–55, 109
  QoS and QoE, 133
  SDN, 134
  self-optimization, 134 (*see also* Self-optimization)
  SONs (*see* Self-organizing networks (SONs))
  3G/4G mobile networks, 133

Z
ZigBee
  application layer framework, 12–13
  automation, 4
  characteristics, 13
  communication protocols, 11
  comparison of IoT, 6
  IoT application, 2
  system architecture, 12
  WPAN technologies, 48

The manufacturer's authorised representative in the EU is Springer Nature Customer Service Centre GmbH, Europaplatz 3, 69115 Heidelberg, Germany. If you have any concerns regarding our products, please contact ProductSafety@springernature.com

Printed and bound by CPI Group (UK) Ltd, Croydon, CR0 4YY
30/07/2026
02181714-0002